Divisor Theory

Harold M. Edwards

Divisor Theory

Springer Science+Business Media, LLC

Harold M. Edwards
Courant Institute of Mathematical Sciences
New York University
New York, New York 10012
U.S.A.

Library of Congress Cataloging-in-Publication Data
Edwards, Harold M.
    Divisor theory / Harold M. Edwards.
       p. cm.

    ISBN 978-0-8176-4976-0   ISBN 978-0-8176-4977-7 (eBook)
    DOI 10.1007/978-0-8176-4977-7
    QA242.E33   1990
    512'.72—dc20   89-28692

Printed on acid-free paper.

© Springer Science+Business Media New York 1990
Originally published by Birkhäuser Boston, in 1990
Softcover reprint of the hardcover 1st edition 1990

Camera-ready text provided by the author using T_EX.
Printed and bound by Edwards Brothers, Inc., Ann Arbor, Michigan.

9 8 7 6 5 4 3 2

Preface

Man sollte weniger danach streben, die Grenzen der mathe-
matischen Wissenschaften zu erweitern, als vielmehr danach,
den bereits vorhandenen Stoff aus umfassenderen Gesichts-
punkten zu betrachten — E. Study

Today most mathematicians who know about Kronecker's
theory of divisors know about it from having read Hermann
Weyl's lectures on algebraic number theory [**We**], and regard
it, as Weyl did, as an alternative to Dedekind's theory of ideals.
Weyl's axiomatization of what he calls "Kronecker's" theory is
built—as Dedekind's theory was built—around unique factorization. However, in presenting the theory in this way, Weyl
overlooks one of Kronecker's most valuable ideas, namely, the
idea that the objective of the theory is to define greatest common divisors, *not* to achieve factorization into primes.

The reason Kronecker gave greatest common divisors the
primary role is simple: they are independent of the ambient
field while factorization into primes is not. The very notion of
primality depends on the field under consideration—a prime
in one field may factor in a larger field—so if the theory is
founded on factorization into primes, extension of the field
entails a completely new theory. Greatest common divisors,
on the other hand, can be defined in a manner that does not
change at all when the field is extended (see §1.16). Only
after he has laid the foundation of the theory of divisors does
Kronecker consider factorization of divisors into divisors prime
in some specified field.

This book gives a full development of a general theory of
divisors (Part 1), together with applications to algebraic number theory (Part 2) and the theory of algebraic curves (Part 3).
There is a preliminary section (Part 0) on a theorem of polynomial algebra that is a natural foundation of the theory, and an
Appendix on differentials, which makes possible the statement

and proof of the Riemann-Roch theorem for curves.

The book began more than ten years ago with an effort to understand Kronecker's theory as it is presented in his treatise *Grundzüge einer arithmetischen Theorie der algebraischen Grössen* (§14 et seq.). (For an early version of the result of these efforts, see the Appendix of [**E1**].) In the intervening years, as my understanding of the theory has increased, I have made many simplifications and extensions of it.

The basic idea is simple: For polynomials with rational coefficients, the content of a product is the product of the contents. (The "content" of a polynomial is the greatest common divisor of its coefficients.) The same *should* be true of polynomials with *algebraic* coefficients. However, the notion of "content" has no obvious meaning for a polynomial with algebraic coefficients. The theory of divisors defines the "content" of a polynomial with algebraic coefficients in such a way that the content of a product is the product of the contents.

In fact, this one requirement determines what the content *must* be in any particular case; the only problems are to show that a consistent theory results and to describe in a simple way what that theory is. As the idea is stated above, and as it will be sketched in this preface, it applies to polynomials whose coefficients are *algebraic numbers* (roots of polynomials in one indeterminate with coefficients in the ring of integers $\mathbf{Z}$) but it applies just as well when the integers are replaced by any natural ring (see §1.2).

Let f and g be polynomials (in any number of indeterminates) whose coefficients are algebraic numbers. There is a polynomial h, whose coefficients are algebraic numbers, such that fh has coefficients in $\mathbf{Z}$. (If the coefficients of f are contained in an algebraic number field K, the norm $N_K f$ of f relative to the field extension $K \supset \mathbf{Q}$ has coefficients in $\mathbf{Q}$, and one can take h to be $(N_K f)/f$ times a common denominator of the coefficients of $N_K f$—see §1.17.) The content of f divides the content of g if and only if the content of fh divides the content of gh (because the contents of fh and gh are the contents of f and g, respectively, times the content of h). But, since fh has coefficients in $\mathbf{Z}$, its content is a positive integer, namely, the greatest common divisor d of the coefficients of fh. Therefore, the statement that the content of fh divides the content of gh has the natural meaning that each coefficient of gh is divisible by d (i.e., the coefficient divided by d is an algebraic integer). Thus, one can *test* whether the content of f divides the content of g by finding an h, determining d, and testing whether the coefficients of gh/d are algebraic integers; the content of f divides the content of g if and only if the answer is yes for all coefficients of gh/d. (See §1.12, Corollary (12).)

The content of a polynomial is, by definition, a *divisor*. As the theory is developed below, the word "content" is not used; the content of f is called "the divisor represented by f." Divisors are represented by polynomials, divisibility of one divisor by another can be tested by the method just described, and two divisors are regarded as *equal* if each divides the other. The task is to show that these definitions result in a consistent theory, and to develop this theory.

The nonzero divisors form a *multiplicative group*. If one specifies an algebraic number field $K \supset \mathbf{Q}$ and restricts consideration to polynomials with coefficients in K, the multiplicative group of divisors coincides with the group of *ideals* in K in the sense of Dedekind. The Kroneckerian theory of divisors has at least three clear advantages over the Dedekindian theory of ideals: (1) It follows from the single, natural premise that the content of the product of two polynomials is the product of the contents. (2) It entails an algorithmic test for divisibility, which, in Dedekindian terms, gives a specific computation for deciding whether a given element is in the ideal generated by a finite set of other elements. (For ideological reasons that are explained in [E2], Dedekind made a *virtue* of the lack of such a test in his theory, whereas Kronecker was of the opposite opinion.) (3) It is independent of the ambient field, so that, unlike the Dedekindian theory, all statements remain true without modification when the ambient field is extended. (The very *definition* of an ideal as a certain kind of subset of the field depends heavily on the ambient field.) Another advantage of the theory of divisors is its applicability to integral and nonintegral divisors alike; the theory of ideals involves both integral and nonintegral ideals, but the integral ideals are far more natural and are usually the only ones introduced in the early stages of the theory.

As was already remarked, the theorem which states that a divisor (or an ideal) can be written in one and only one way as a product of powers of distinct prime divisors (ideals) has meaning only when an ambient field is specified, because only then does the word "prime" have meaning. However, many applications of this theorem require only a decomposition into a product of powers of *relatively* prime divisors, a notion independent of the ambient field. Thus, factorization into primes can often be replaced by Theorem 1 of §1.19, which states that the divisors in any given finite set can be written as products of powers of relatively prime integral divisors.

A second fundamental theorem of divisor theory is the following: An integral divisor divides at least one algebraic integer. Therefore, given an integral divisor A, there is an integral divisor B such that AB is the divisor of an algebraic integer. Theorem 2 (§1.20) states that, given any integral divisor C, one can in fact choose B to be relatively prime to C (i.e., there is an integral divisor B relatively prime to C for which AB is the divisor of an algebraic integer). One corollary of this theorem is that every divisor is the greatest common divisor of just *two* algebraic numbers. Another corollary is a divisibility test of the type used by Kummer in his definition of "ideal prime factors" of cyclotomic integers, the notion with which he initiated divisor theory, in a special case, in 1846. (See also §2.11.)

The remaining topics treated in the general theory of Part 1 all relate to divisors in some particular ambient field. They include divisor class groups, the factorization of prime divisors in normal extension fields, rings of values of a given field at a given integral divisor, discriminants, differents, and ramification. A topic *not* covered in the general theory is the representation of an arbitrary divisor as a product of powers of divisors prime in a given field. The problem of giving, in the general case, an algorithm for factoring an integral divisor into irreducible (prime) factors (see §1.18) appears to be more difficult than the problems treated here. At any rate, it is a problem for which I do not have a solution.

However, in the special case of Part 2—the case of algebraic number fields—it is quite simple to write any given divisor as a product of powers of prime divisors (§2.1). Much of Part 2 is devoted to proving the validity of the following method for factoring the divisor of a prime integer p in an algebraic number field. Let α_1, α_2, ..., α_ν be algebraic integers and let $K = \mathbf{Q}(\alpha_1, \alpha_2, \ldots, \alpha_\nu)$ be the field they generate over $\mathbf{Q}$. Let X, u_1, u_2, ..., u_ν be indeterminates, let $F(X, u_1, u_2, \ldots, u_\nu)$ be the norm of $X - u_1\alpha_1 - u_2\alpha_2 - \cdots - u_\nu\alpha_\nu$ relative to $K \supset \mathbf{Q}$ (F is a polynomial with coefficients in $\mathbf{Z}$), and let $F \equiv f_1^{e_1} f_2^{e_2} \cdots f_m^{e_m} \bmod p$ be the factorization of F mod p into powers of distinct irreducible polynomials with coefficients in the field $\mathbf{Z}$ mod p. *Normally* the divisor P_i represented by the polynomial $p + f_i(\alpha_1 u_1 + \cdots + \alpha_\nu u_\nu, u_1, u_2, \ldots, u_\nu)$ is prime in K, $P_i \neq P_j$ for $i \neq j$, and $P_1^{e_1} P_2^{e_2} \cdots P_m^{e_m}$ is the divisor of p. For a given set of α's, this is true for all but a finite number of primes p (§2.4). If the α's have the property that every algebraic integer in K can be expressed as a polynomial in the α's with coefficients in $\mathbf{Z}$—and for given K such a set of α's can always be found—it is true for *all* primes p.

This method of factoring the divisor of p was first proposed by Kronecker [**Kr1**, §25], who proposed it in a much more general case than the case of algebraic number fields. The va-

lidity of the method in the number field case was proved by
Hensel, but, although Hensel stated [**He**, p. 76] that he had a
proof which he would soon publish, I do not know of a proof
of Kronecker's more general case. (If one could be given, it
would be a large step toward the solution of the problem of
factoring divisors in the general case.)

Two other topics treated in Part 2 are Dedekind's discriminant theorem (§2.8) and the factorization of primes in subfields
of cyclotomic fields (§2.11).

Application of the theory of divisors to algebraic curves in
Part 3 calls for a slight extension of the notion of "divisor."
The field of functions K on an algebraic curve over the ratio-
nals is an algebraic extension of the natural ring $\mathbf{Q}[x]$ and as
such has a divisor theory. This divisor theory depends, however, on the choice of a parameter x on the curve. A *global*
divisor in such a field K is the assignment to each parameter
x of a divisor A_x in the divisor theory for this parameter in
such a way that the divisors A_x "agree on overlaps" in a natural way (§3.4). In Part 3, unless otherwise stated, "divisor"
means "global divisor."

Divisor theory provides the following answer to the perennial
question "What is a point?" in the theory of algebraic curves.
Let K be the field of rational functions on the algebraic curve
defined by the equation $F(x, y) = 0$, where F is an irreducible
polynomial in two indeterminates with coefficients in $\mathbf{Z}$. Let
a and b be algebraic numbers in K such that $F(a, b) = 0$.
There is associated to such a pair of algebraic numbers a divisor
in K, namely, the numerator of the divisor represented by
$(x - a)U + (y - b)V$ (where $x - a$ and $y - b$ are elements of
K, and U and V are indeterminates). If (a, b) is a *nonsingular*
point of $F = 0$, that is, if the partial derivatives of F at (a, b)
are not both zero, the divisor in K obtained in this way is
called a *place* (see §3.13). Places can also be characterized as
divisors in K which are prime in K and in all extensions of
K obtained by adjoining constants to K (§3.22). Yet another

characterization of places is as divisors for which the ring of values coincides with the field of constants of K.

Places capture algebraically the notion of a point on a curve. For example, the origin $(0,0)$ does not give rise to a place on the folium of Descartes $x^3 + y^3 - xy = 0$, but, rather, gives rise to a *product of two* places, namely, the zero and the pole of the "function" x/y on this curve. That the origin is a product of two places expresses the geometrical "fact" that the origin is a double point of the folium of Descartes (§3.13).

A fundamental theorem states that every divisor in the field of functions K on an algebraic curve, can, when suitable constants are adjoined, be written as a product of powers of places (§3.18). (This theorem relates divisors as they are treated in this book to divisors as they are customarily defined in the theory of algebraic curves as formal sums of places with integer coefficients, or, equivalently, as formal products of places with integer exponents.) The *degree* of a divisor is equal to the number of places in the numerator of such a representation of the divisor minus the number of places in the denominator. The degree of the divisor of an element x of K is always 0, and the places in its numerator are naturally thought of as "the points where x is zero" and the places in its denominator as "the points where x has poles."

Divisor theory also provides the following natural formulation of Abel's theorem: For any given function field (of one variable, over $\mathbf{Q}$) there is a least integer g, the genus of the field, such that every divisor of degree g is equivalent to an integral divisor. Otherwise stated, given a set of zeros and a set of poles, there is a function on the curve with poles, at most, at the given poles and zeros at the given zeros (plus, possibly, other zeros) provided the number of given poles is at least g greater than the number of given zeros. This theorem is not immediately recognizable as Abel's theorem, but the connection with Abel's own statement is explained in §3.25. (See also the Corollary of §3.27.)

An element of K can be expanded in a natural way in powers of a local parameter at a place P in K (§3.16). The *principal part* of an element of K at P, relative to a given local parameter at P, is the terms of this power series expansion which have negative degree. (In particular, the element of K has a pole at P if and only if its principal part at P, relative to any local parameter at P, is nonzero.) Let P_1, P_2, ..., P_k be a given set of places in the field K of rational functions on an algebraic curve, and let Γ be the elements of K which have poles only at the P_i and whose poles at the P_i have order N at most. The principal parts of an element of Γ are described by Nk constants of K. Abel's theorem easily implies that the principal parts of elements of Γ form a subspace of codimension at most g of K_0^{Nk}. The Appendix is devoted to proving that *this subspace of K_0^{Nk} can be described in terms of differentials.* Specifically, differentials are defined in the Appendix, and it is shown that (1) g is the dimension of the vector space (over the field of constants K_0 of K) of holomorphic differentials, (2) the sum of the residues of any differential is zero, and (3) the conditions "any element of K times any holomorphic differential must have the sum of its residues equal to zero" give necessary and sufficient conditions for determining which elements of K_0^{Nk} are principal parts of elements of Γ. (For N large, these conditions are also independent, so that the codimension is exactly g when N is large.) The Riemann-Roch theorem is a simple corollary of this method of determining the principal parts of elements of Γ.

I wish once again to thank the Vaughn Foundation for more than a decade of support which made an enormous difference in my life and work.

Contents

Contents

Part 0: A Theorem of Polynomial Algebra

What is usually called "Gauss's lemma" today differs from Gauss's actual lemma (*Disquisitiones Arithmeticae*, Art. 42).

GAUSS'S VERSION. *Let f and g be monic polynomials in one indeterminate with rational coefficients. If the coefficients of f and g are not all integers, then the coefficients of fg cannot all be integers.*

MODERN VERSION. (cf. Bourbaki, *Algèbre Commutative*, VII, 3, 5) *Let f and g be polynomials in one indeterminate with integer coefficients. The content of fg is the content of f times the content of g.*

(The "content" of a polynomial with integer coefficients is the greatest common divisor of its coefficients.) Gauss's version has the advantage that it lends itself to the following profound generalization.

GENERALIZATION. *Let f and g be monic polynomials in one indeterminate whose coefficients are algebraic numbers. If the coefficients of f and g are not all algebraic integers, then the coefficients of fg cannot all be algebraic integers.*

(An algebraic number is an element of an extension of finite degree of the field of rational numbers $\mathbf{Q}$. The coefficients of f and g are assumed to be elements of a single extension field K of $\mathbf{Q}$ of finite degree. An element of K is an "algebraic integer" if some power of it can be written as a linear combination of lower powers with integer coefficients.)

The modern version has no such generalization because the notion of "greatest common divisor" is not defined for algebraic integers. On the other hand, the modern version of Gauss's lemma does have advantages—for example, it is no harder to prove, and it easily implies Gauss's version, but Gauss's version does not easily imply it—so a formulation of the modern version which generalizes to the case of algebraic numbers in the way that Gauss's does would be desirable. Dedekind recognized this need and published such a formulation [**D1**] in 1892.

DEDEKIND'S VERSION. *Let f and g be polynomials in one indeterminate with rational coefficients. If fg has integer coefficients, then the product of any coefficient of f and any coefficient of g is an integer.*

This version is as easy to prove as Gauss's, and it implies both Gauss's version and the modern version. (Since 1 is a coefficient of a monic polynomial, Gauss's version follows immediately. Let f and g have integer coefficients and let a and b, respectively, be their contents. Since ab divides all coefficients of fg, it divides the content of fg. If it were not equal to the content of fg, there would be a prime p such that the coefficients of $(fg)/(abp)$ were integers. It would then follow from Dedekind's version of Gauss's lemma that every coefficient of f/a times every coefficient of g/b was divisible by p, which is impossible because both f/a and g/b have a coefficient not divisible by p, and p is prime.)

Dedekind's version does generalize, as he proved, to the algebraic case. He published this theorem in the Proceedings of the German Mathematical Society of Prague and later referred to it as the Prague theorem.

DEDEKIND'S PRAGUE THEOREM. *Let f and g be polynomials in one indeterminate whose coefficients are algebraic numbers. If all coefficients of fg are algebraic integers, then the product of any coefficient of f and any coefficient of g is an algebraic integer.*

Dedekind was understandably not aware that his Prague theorem was a consequence of a theorem published ten years earlier by Kronecker [**Kr2**]. Kronecker's brief paper is exceedingly terse and obscure. However, it contains, couched in Kronecker's private terminology, a theorem which implies:

THEOREM. *Let $a_0, a_1, \ldots, a_m, b_0, b_1, \ldots, b_n$ be indeterminates and let R be the ring of polynomials in these indeterminates with integer coefficients. Let $c_0, c_1, \ldots, c_{m+n} \in R$ be defined by*

$$c_i = \sum_{\substack{j+k=i \\ 0 \le j \le m, 0 \le k \le n}} a_j b_k.$$

(In other words, c_i is the coefficient of x^{m+n-i} in $(a_0 x^m + a_1 x^{m-1} + \cdots + a_m)(b_0 x^n + b_1 x^{n-1} + \cdots + b_n)$, where x is a new indeterminate.) Then each of the $(m+1)(n+1)$ elements $a_j b_k$ $(0 \le j \le m, 0 \le k \le n)$ of R is integral over the subring of R generated by $1, c_0, c_1, \ldots, c_{m+n}$. In other words, some power of each $a_j b_k$ can be written as a linear combination of lower powers

$$(a_j b_k)^N = p_1 \cdot (a_j b_k)^{N-1} + p_2 \cdot (a_j b_k)^{N-2} + \cdots + p_N$$

in which the coefficients p_i are elements of R expressible as polynomials in $c_0, c_1, \ldots, c_{m+n}$ with integer coefficients.

Examples: When $m = n = 1$,

$$(a_0 b_1)^2 = c_1 (a_0 b_1) - c_0 c_2,$$

as is easily verified by substituting $c_0 = a_0 b_0, c_1 = a_0 b_1 + a_1 b_0, c_2 = a_1 b_1$.

When $m = n = 2$,

$$(a_0 b_1)^6 = p_1 (a_0 b_1)^5 + p_2 (a_0 b_1)^4 + \cdots + p_6$$

where

$$p_1 = 3c_1, \quad p_2 = -3c_1^2 - 2c_0 c_2, \quad p_3 = c_1^3 + 4c_0 c_1 c_2,$$

$$p_4 = -c_0^2 c_1 c_3 - 2c_0 c_1^2 c_2 - c_0^2 c_2^2 + 4c_0^3 c_4,$$

$$p_5 = c_0^2 c_1^2 c_3 + c_0^2 c_1 c_2^2 - 4c_0^3 c_1 c_4, \quad p_6 = -c_0^3 c_1 c_2 c_3 + c_0^4 c_3^2 + c_0^3 c_1^2 c_4$$

(Found with the aid of a computer.)

COROLLARY 1. *Let f and g be polynomials in one indeterminate with coefficients which are algebraic numbers, and let fg have integer coefficients. Then any coefficient of f times any coefficient of g is an algebraic integer.*

DEDUCTION: Substitute the coefficients of f and g in the obvious way into the formula $(a_j b_k)^N = \sum p_i \cdot (a_j b_k)^{N-i}$ of the theorem. Since this substitution gives the c's integer values, it gives the p's integer values and shows that the value it gives to $a_j b_k$ is an algebraic integer.

Corollary 1 is a weakened version of Dedekind's Prague theorem—it applies only when fg has integer coefficients, not, as Dedekind's theorem does, when fg has algebraic integer coefficients. It is easy to deduce Dedekind's theorem itself as a corollary of the theorem (see Proposition 2 below), but since Dedekind's theorem will not be needed in the sequel, this deduction will be left to the reader.

COROLLARY 2. *Dedekind's version of Gauss's lemma.*

DEDUCTION: This follows from Corollary 1 and the elementary fact that a rational number which is an algebraic integer is an integer.

COROLLARY 3. *Corollary 1 is also true for polynomials in any number of indeterminates. That is, if f and g are polynomials (in any number of indeterminates) whose coefficients are algebraic numbers, and if fg has integer coefficients, then the product of any coefficient of f and any coefficient of g is an algebraic integer.*

DEDUCTION: Let $X_1, X_2, \ldots, X_\nu$ be the indeterminates which appear in f and g, and let k be an integer greater than the degree of fg in any of the indeterminates X_i it contains. Define a homomorphism ϕ from polynomials in $X_1, X_2, \ldots, X_\nu$ to polynomials in T by $\phi(X_i) = T^{k^{i-1}}$. Because $e_1 + e_2 k + e_3 k^2 + \cdots = e_1' + e_2' k + e_3' k^2 + \cdots$ for $0 \le e_i < k$, $0 \le e_i' < k$ implies

$e_i = e_i'$, ϕ applied to different monomials $X_1^{e_1} X_2^{e_2} \ldots X_\nu^{e_\nu}$ in which the exponents e_i are all less than k gives different powers of T. Therefore, the coefficients of $\phi(fg) = \phi(f)\phi(g)$ *coincide with the coefficients of* fg. By Corollary 1, any coefficient of $\phi(f)$ times any coefficient of $\phi(g)$ is an algebraic integer. Since the coefficients of $\phi(f)$ coincide with those of f, and of $\phi(g)$ with those of g, Corollary 3 follows.

COROLLARY 4. *Let r be an integral domain and let k be its field of quotients. If f and g are polynomials with coefficients in an algebraic extension K of k such that fg has coefficients in r, then the product of any coefficient of f and any coefficient of g is integral over r (that is, some power of it is equal to a linear combination of lower powers with coefficients in r).*

DEDUCTION: The case where f and g are polynomials in one indeterminate follows from the Theorem as in Corollary 1. The case where they are polynomials in several indeterminants then follows as in Corollary 3.

It was noted above that the modern version of Gauss's lemma does not apply to polynomials whose coefficients are algebraic integers because the concept of "greatest common divisor" is not defined for sets of algebraic integers. The basic idea of the theory of divisors is, roughly, to use this identity "content of product = product of contents" to *define* the content of a polynomial whose coefficients are algebraic numbers. As is shown in Part 1, the key to proving that this idea leads to a coherent theory with all the important characteristics of "greatest common divisors" is the above theorem—and more particularly Corollary 4. Kronecker was unquestionably aware of the connection, because he opens his paper with a mention of "the effort to find the simplest foundation of the theory of forms" (his major treatise [Kr1], published one year earlier, in which he develops the theory of divisors, treats divisors as an aspect of the theory of "forms" or polynomials). However, Kronecker appears never to have published anything further on the subject of "the simplest foundation of the theory of forms", and the account in this one paper is, as I have already said, opaque.

PROOF OF THE THEOREM:(Hurwitz[**Hu2**]) Let $x, t_0, t_1, \ldots, t_n$ be indeterminates, and let $\widehat{R} = R[x, t_0, t_1, \ldots, t_n] = \mathbf{Z}[a_0, \ldots, a_m, b_0, b_1, \ldots, b_n, x, t_0, t_1, \ldots, t_n]$ be the ring of polynomials in all these indeterminates with coefficients in the ring of integers $\mathbf{Z}$. Let $f(x) = a_0 x^m + a_1 x^{m-1} + \cdots + a_m \in \widehat{R}$ and $g(x) = b_0 x^n + b_1 x^{n-1} + \cdots + b_n \in \widehat{R}$.

LEMMA 1. (Lagrange Interpolation) In $\widehat{R}$,

$$g(x) \prod_{j \neq k} (t_j - t_k) = \sum_{i=0}^{n} Q_i(t_0, t_1, \ldots, t_n) g(t_i) \prod_{j \neq i} (x - t_j)$$

where the product on the left is over all $(n+1)^2 - (n+1)$ factors $t_j - t_k$ in which $0 \leq j \leq n$, $0 \leq k \leq n$, $j \neq k$, where $g(t_i) = \sum_{j=0}^{n} b_j t_i^{n-j} \in \widehat{R}$, where the products on the right are over all n factors $x - t_j$ in which $0 \leq j \leq n$, $j \neq i$, and where

$$Q_i(t_0, t_1, \ldots, t_n) = \frac{\prod_{j \neq k} (t_j - t_k)}{\prod_{j \neq i} (t_i - t_j)},$$

a polynomial in the t's of degree $(n+1)^2 - (n+1) - n = n^2$.

PROOF: Both sides are elements of $\widehat{R}$ of degree n, at most, in x, so their difference is also a polynomial of degree n, at most, in x. This difference is divisible by $x - t_\nu$ for $\nu = 0, 1, \ldots, n$ by the Remainder Theorem, because division of the left side by $x - t_\nu$ leaves a remainder of $g(t_\nu) \prod_{j \neq k} (t_j - t_k)$ and division by $x - t_\nu$ of the one term of the sum on the right side which is not divisible by $x - t_\nu$ leaves the same remainder $Q_\nu(t_0, t_1, \ldots, t_n) g(t_\nu) \prod_{j \neq \nu} (t_\nu - t_j) = g(t_\nu) \prod_{j \neq k} (t_j - t_k)$. Thus, the difference is divisible by the polynomial $\prod_{j=0}^{n} (x - t_j)$ of degree $n + 1$ in x, which implies that the difference is zero.

LEMMA 2. *Let S denote the subset of R consisting of those elements which can be expressed as polynomials in the a's and c's that are homogeneous of degree n in the a's and homogeneous of degree 1 in the c's. The coefficients of $g(x) \prod_{i=0}^{n} f(t_i) \prod_{j \neq k}(t_j - t_k)$, regarded as a polynomial in x and the t's with coefficients in R, are in S.*

PROOF: Multiplication of the formula of Lemma 1 by $\prod_{i=0}^{n} f(t_i)$ shows that the polynomial in question is equal to

$$\sum_{i=0}^{n} Q_i(t_0, t_1, \ldots, t_n) f(t_i) g(t_i) \prod_{j \neq i} f(t_j) \prod_{j \neq i}(x - t_j).$$

Since $f(t_i)g(t_i) = \sum_{\nu=0}^{m+n} c_\nu t_i^{m+n-\nu}$ and $f(t_j) = \sum_{\nu=0}^{m} a_\nu t_j^{m-\nu}$, while the Q_i and $\prod_{j \neq i}(x - t_j)$ have coefficients in $\mathbf{Z}$, the lemma follows.

LEMMA 3. *The coefficients of $g(x) \prod_{i=0}^{n} f(t_i)$, regarded as a polynomial in x and the t's with coefficients in R, are in S.*

PROOF: Let the polynomial of Lemma 2 be successively divided by the $(n+1)^2 - (n+1)$ polynomials $t_j - t_k$. As follows from the division algorithm, the coefficients of each successive quotient are in S. Explicitly, if $(t_j - t_k)H = G$, and if the coefficients of G are in S, then the coefficients of H are in S because, when $G = G_0 + G_1 t_j + G_2 t_j^2 + \cdots + G_\mu t_j^\mu$ is divided by $t_j - t_k$ to find $H = H_0 + H_1 t_j + \cdots + H_{\mu-1} t_j^{\mu-1}$, the division algorithm gives $H_{\mu-1} = G_\mu$, $H_{\mu-2} = G_{\mu-1} + t_k G_\mu$, $H_{\mu-3} = G_{\mu-2} + t_k G_{\mu-1} + t_k^2 G_\mu$, $\ldots$. Therefore, if all coefficients of G are in S, so are all coefficients of H. The initial polynomial is the polynomial of Lemma 2, and the final one is the polynomial of Lemma 3.

LEMMA 4. *Let M_1, M_2, $\ldots$, M_τ be a list of all monomials $a_0^{e_0} a_1^{e_1} \cdots a_m^{e_m}$ in the a's of degree n. The equations $a_j b_k M_\rho = \sum_{\sigma=1}^{\tau} \lambda_{jk\rho\sigma} M_\sigma$ define linear polynomials $\lambda_{jk\rho\sigma}$ in*

$c_0, c_1, \ldots, c_{m+n}$ with coefficients in $\mathbf{Z}$ for all sets of indices in the ranges $0 \le j \le m$, $0 \le k \le n$, $0 \le \rho \le \tau$, $0 \le \sigma \le \tau$.

PROOF: Each $a_j b_k M_\rho$ is a coefficient of $g(x) \prod_{i=0}^n f(t_i)$. Therefore, it has the desired form by Lemma 3.

CONCLUSION OF THE PROOF OF THE THEOREM: For fixed indices j and k, $0 \le j \le m$, $0 \le k \le n$, let Λ_{jk} be the $\tau \times \tau$ matrix whose entries are the $\lambda_{jk\rho\sigma}$ of Lemma 4. Then $(a_j b_k I - \Lambda_{jk})M = 0$, where I is the $\tau \times \tau$ identity matrix, where M is the column matrix with entries M_1, M_2, $\ldots$, M_τ, and where the 0 on the right represents a column matrix of length τ whose entries are all 0. Since M is nonzero with entries in the integral domain R and $a_j b_k I - \Lambda_{jk}$ is a square matrix with entries in R, it follows that $\det(a_j b_k I - \Lambda_{jk}) = 0$. In other words, $a_j b_k$ is a root of the polynomial $\det(XI - \Lambda_{jk})$ in the indeterminate X. Since this polynomial has the form $X^\tau + p_1 X^{\tau-1} + p_2 X^{\tau-2} + \cdots + p_\tau$, where the p_i are polynomials in the λ's, the Theorem follows.

An equivalent theorem somewhat closer to the theorem Kronecker states is:

ALTERNATE THEOREM. *As before, let* $R = \mathbf{Z}[a_0, a_1, \ldots, a_m, b_0, b_1, \ldots, b_n]$ *and let* $c_i = \sum_{j+k=i} a_j b_k$. *Let* G *be* $(a_0 x_0 + a_1 x_1 + \cdots + a_m x_m)(b_0 y_0 + b_1 y_1 + \cdots + b_n y_n)$, *a polynomial in* $x_0, x_1, \ldots, x_m, y_0, y_1, \ldots, y_n$ *with coefficients in* R. *Then, for some positive integer* ρ, G^ρ *can be expressed as a linear combination of lower powers of* G *with coefficients which are polynomials in the* c's, x's, *and* y's *with coefficients in* $\mathbf{Z}$.

Example: (Kronecker [**Kr2**, p. 424]) When $m = n = 1$, $G^2 - PG + Q = 0$ where, with $G' = (b_0 x_0 + b_1 x_1)(a_0 y_0 + a_1 y_1)$, $P = G + G'$ and $Q = GG'$. That P and Q are polynomials in $c_0, c_1, c_2, x_0, x_1, y_0, y_1$ can be verified as follows. $G = c_0 x_0 y_0 + a_0 b_1 x_0 y_1 + a_1 b_0 x_1 y_0 + c_2 x_1 y_1$. Thus $P = 2c_0 x_0 y_0 + c_1(x_0 y_1 + x_1 y_0) + 2c_2 x_1 y_1$ and $Q = (c_0 x_0 y_0 + c_2 x_1 y_1)^2 + (c_0 x_0 y_0 + c_2 x_1 y_1) \cdot (c_1 x_0 y_1 + c_1 x_1 y_0) + (a_0 b_1 x_0 y_1 + a_1 b_0 x_1 y_0)(a_1 b_0 x_0 y_1 + a_0 b_1 x_1 y_0)$;

the last term in this expression of Q is $c_0 c_2 x_0^2 y_1^2 + (a_0^2 b_1^2 + a_1^2 b_0^2) x_0 x_1 y_0 y_1 + c_0 c_2 x_1^2 y_0^2$, which can be expressed in terms of c's by virtue of $c_1^2 = a_0^2 b_1^2 + a_1^2 b_0^2 + 2 c_0 c_2$.

The previous theorem follows from this one when 1 is substituted for one x and one y, and 0 is substituted for all the others. Conversely, this theorem follows from the previous one when use is made of a few basic facts about integrality:

DEFINITION. An element a of a ring R is *integral* over a subring R' of R if some power of a is equal to a linear combination of lower powers with coefficients in R'.

PROPOSITION 1. *An element a of R is integral over a subring R' of R if and only if there exists, for some positive integer ν, a set of elements $\omega_1, \omega_2, \ldots, \omega_\nu$ of R, not all zero, such that $a \omega_i = \lambda_{i1} \omega_1 + \lambda_{i2} \omega_2 + \cdots + \lambda_{i\nu} \omega_\nu$ for some $\nu \times \nu$ matrix of elements $\lambda_{ij} \in R'$.*

PROOF: As in the proof of the theorem above, the equations $a \omega_i = \sum \lambda_{ij} \omega_j$ imply, when the ω's are not all zero, that a is a root of $\det(XI - \Lambda)$, where I is the $\nu \times \nu$ identity matrix, where $\Lambda = (\lambda_{ij})$, and where X is an indeterminate. Therefore, since $\det(XI - \Lambda) = X^\nu + p_1 X^{\nu-1} + p_2 X^{\nu-2} + \cdots$ where the p_i are in R', a is integral over R'. Conversely, if a is integral over R', say $a^\mu = q_1 a^{\mu-1} + q_2 a^{\mu-2} + \cdots + q_\mu$ with $q_i \in R'$, then, with $\omega_i = a^{i-1}$ for $i = 1, 2, \ldots, \mu$, $a \omega_i = \omega_{i+1}$ for $i = 1, 2, \ldots, \mu - 1$, and $a \omega_\mu = q_1 \omega_{\mu-1} + q_2 \omega_{\mu-2} + \cdots + q_\mu \omega_1$.

COROLLARY. *Let R' be a subring of a ring R. The elements of R integral over R' are a subring of R containing R'.*

DEDUCTION: An element a of R' is integral over R' by virtue of $a = a \cdot a^0$. If a and b are elements of R integral over R', then there exist nonzero elements $\omega_1, \omega_2, \ldots, \omega_\nu, \omega_1', \omega_2', \ldots, \omega_\mu'$ such that $a \omega_i = \sum_{j=1}^\nu \lambda_{ij} \omega_j$ and $b \omega_i' = \sum_{j=1}^\mu \lambda_{ij}' \omega_j'$, where the λ_{ij} and λ_{ij}' are in R'. The $\mu\nu$ elements $\omega_\sigma \omega_\tau'$ ($1 \le \sigma \le \nu$, $1 \le \tau \le \mu$) satisfy $a \omega_\sigma \omega_\tau' = \sum_{j=1}^\nu \lambda_{\sigma j} \omega_j \omega_\tau'$ and $b \omega_\sigma \omega_\tau' =$

$\sum_{j=1}^{\mu} \lambda'_{\tau j} \omega_\sigma \omega'_j$, from which it follows easily that both $(a + b)\omega_\sigma \omega'_\tau$ and $ab\omega_\sigma \omega'_\tau$ can be expressed as linear combinations of the $\mu\nu$ elements $\omega_\sigma \omega'_\tau$ with coefficients in R'. Therefore, the elements of R integral over R' are closed under addition and multiplication.

PROOF OF THE ALTERNATE THEOREM: $G = \sum\sum a_j b_k x_j y_k$. Let R' be the subring of the ring of polynomials with integer coefficients in the a's, b's, x's, and y's which is generated by the c's, the x's, and the y's. Then the x's and y's are integral over R'. By the theorem already proved, each $a_j b_k$ is integral over R'. Therefore, G is integral over R' by the Corollary.

Two postscripts: 1) Another basic proposition about integrality will be needed.

PROPOSITION 2. *Let R' be a subring of a ring R, and let R'' be the ring of elements of R integral over R'. Any element of R integral over R'' is in R''.*

PROOF: Let $b \in R$ be integral over R'', say $b^n = a_1 b^{n-1} + a_2 b^{n-2} + \cdots + a_n$ where the $a_i \in R''$. For each $i = 1, 2, \ldots, n$ let $a_i^N = \sum_{j=1}^{N} \lambda_{ij} a_i^{N-j}$ where $\lambda_{ij} \in R'$. (There is clearly no loss of generality in using the same exponent N for all i.) Let the ω's be the nN^n products $b^e a_1^{e_1} a_2^{e_2} \cdots a_n^{e_n}$ in which $0 \le e < n$ and $0 \le e_i < N$. It is easy to see that b times any one of these nN^n elements ω is a linear combination of ω's with coefficients in R'. Therefore, b is integral over R', as was to be shown.

2) Kronecker's proof of the theorem, while less slick than the proof just given, sheds more light on the reason the theorem is true. Briefly, the argument Kronecker sketches* is as follows:

Let $R_1 = \mathbf{Z}[a_0, b_0, u_1, u_2, \ldots, u_{m+n}]$ be the ring of polynomials in $m + n + 2$ indeterminates with these names (integer

---

*Kronecker's proof [Kr2] is not at all clear. I am grateful to Dr. Olaf Neumann of Jena for explaining it to me. See [E-N-P, p. 69]

coefficients). Define $a_1, a_2, \ldots, a_m, b_1, b_2, \ldots, b_n, c_0, c_1, \ldots,$ $c_{m+n} \in R_1$ by the equations

$$\sum_{i=0}^{m} a_i x^{m-i} = a_0 \prod_{i=1}^{m} (1 + u_i x)$$

$$\sum_{i=0}^{n} b_i x^{n-i} = b_0 \prod_{i=1}^{n} (1 + u_{m+i} x)$$

$$\sum_{i=0}^{m+n} c_i x^{m+n-i} = a_0 b_0 \prod_{i=1}^{m+n} (1 + u_i x).$$

(In other words, a_i is a_0 times the ith elementary symmetric function in the first m of the u's, b_i is b_0 times the ith elementary symmetric function in the last n of the u's, and c_i is $a_0 b_0$ times the ith elementary symmetric function in the u's.) Let G be $\sum\sum a_j b_k x_j y_k$, a polynomial in new indeterminates $x_0, x_1, \ldots, x_m, y_0, y_1, \ldots, y_n$ with coefficients in R_1. Let $\rho = (m + n)!$. The ρ permutations of the u's act on R_1 and therefore act on polynomials with coefficients in R_1. These permutations produce ρ conjugates of G, call them $G_1, G_2, \ldots, G_\rho$, all of them polynomials in the x's and y's with coefficients in R_1. Let $F(X) = \prod_{i=1}^{\rho}(X - G_i)$ where X is a new indeterminate. Then $F(G) = 0$ (G is among the G_i) and the coefficient of $X^{\rho-r}$ in F is the rth elementary symmetric function in $G_1, G_2, \ldots, G_\rho$, call it S_r. The main step is to show that S_r can be expressed as a polynomial in the x's, y's, and c's with integer coefficients.

Let S_r be regarded as a polynomial in the x's and y's with coefficients in R_1, and let C be one of its coefficients. Since each coefficient of G is $a_0 b_0$ times a polynomial in the u's which has degree 1 in any one u, C is $a_0^r b_0^r$ times a *symmetric* polynomial in the u's which has degree r, at most, in any one u. By the fundamental theorem on symmetric polynomials, C is $a_0^r b_0^r$ times a polynomial in $c_1/c_0, c_2/c_0, \ldots, c_{m+n}/c_0,$

say $C = a_0^r b_0^r \cdot p(c_1/c_0, c_2/c_0, \ldots, c_{m+n}/c_0)$, where p has integer coefficients. Since $a_0 b_0 = c_0$, the factor c_0^r clears the denominators of $p(c_1/c_0, c_2/c_0, \ldots, c_{m+n}/c_0)$—so that C is a polynomial in the c's—by virtue of:

LEMMA. *Let $h(X_1, X_2, \ldots, X_\mu)$ be a polynomial of total degree ν with integer coefficients. Let $z_1, z_2, \ldots, z_\mu$ be indeterminates, and let $\sigma_1, \sigma_2, \ldots, \sigma_\mu$ be the elementary symmetric functions in the z's. Then $h(\sigma_1, \sigma_2, \ldots, \sigma_\mu)$ has degree ν in each z.*

The lemma can be proved by writing all polynomials using lexicographic order and examining the leading terms. The leading term of $h(\sigma_1, \sigma_2, \ldots, \sigma_\mu)$ has degree ν in z_1, so by symmetry its degree in all z's is ν.

Thus $F(G) = 0$ is a relation of the desired form, except that the a's and b's it relates are contained in R_1 and are not the indeterminates one begins with. To prove the theorem, one must show that the mapping $R \to R_1$ is one-to-one, so that the relation $F(G) = 0$ in $R_1[x_0, x_1, \ldots, x_m, y_0, y_1, \ldots, y_n]$ implies a relation in $R[x_0, x_1, \ldots, x_m, y_0, y_1, \ldots, y_n]$ of the same form.

Dedekind gave two other proofs of the theorem [D1], each containing interesting ideas.

Part 1: The General Theory

§1.1 Introduction.

The general theory of divisors, as it is developed in this first part, applies to algebraic extension fields of rings that are *natural* in the sense defined in the next article. The case of number theory in Part 2 is the case of the natural ring $\mathbf{Z}$, and the case of algebraic curves in Part 3 is the case of the natural ring $\mathbf{Q}[x]$.

The objective of the theory is to define divisors and divisibility of elements by divisors in such a way that, given a natural ring r and given a (finite) set of elements, all of which are algebraic over r, there is a divisor which is their *greatest common divisor*. Divisors form a multiplicative *group* in which there are *integral* elements (a divisor is integral if it is the greatest common divisor of a set of elements all of which are integral over the ring) and in which there is consequently a notion of *divisibility* (A divides B if B/A is integral). To say that a divisor A is the greatest common divisor of the set of elements α_1, α_2, ..., α_ν algebraic over r means simply that A divides α_i for all i and is divisible by any divisor which divides α_i for all i.

The core of the theory is presented, after a few preliminaries, in §1.5–§1.12. The remainder of Part 1 is devoted to various developments of the theory that can be carried out in the case of a general natural ring.

§1.2 Natural rings.

The theory of divisors applies to quantities algebraic over what I will call "natural" rings.

DEFINITION. A ring r is *natural* if

(1) r is a commutative integral domain with multiplicative identity 1.

(2) r is a g.c.d. domain. That is, given any two nonzero

elements $a, b \in r$, one can find a third element $c \in r$ which divides both a and b and is divisible by any $d \in r$ which divides both a and b.

(3) Factorizations of nonzero elements of r contain a finite number of factors. More precisely, given $a \in r$, $a \neq 0$, one can find a positive integer n such that any factorization of a into n factors in r contains at least one factor which is a unit. (An element of r is a unit if it divides 1, or, what is the same, if it divides all elements of r.)

(4) There is a factorization method for polynomials in one variable with coefficients in r. That is, given a polynomial $f(x) = a_0 x^n + a_1 x^{n-1} + \cdots + a_n$ in one indeterminate x with coefficients a_i in r, a finite calculation produces *either* a factorization $f = gh$ of f into two factors g and h, neither of which is a unit, *or* a proof that there is no such factorization. (A polynomial g in x with coefficients in r is a unit if, as a polynomial, it divides 1, which is the case if and only if it is a constant polynomial whose only term is a unit of r.)

Kronecker applied his theory to rings of the form $r = \mathbf{Z}[x_1, x_2, \ldots, x_n]$ of polynomials in some number of indeterminates—possibly none—with coefficients in the ring of integers $\mathbf{Z}$. It is elementary to prove that these rings, which Kronecker called "natural", are natural in the sense just defined (see, for example, [**E3**]). In addition, rings of polynomials $r = \mathbf{Q}[x_1, x_2, \ldots, x_n]$ with coefficients in the rational field $\mathbf{Q}$ or rings of polynomials $r = \mathbf{F}_q[x_1, x_2, \ldots, x_n]$ with coefficients in a finite field $\mathbf{F}_q$ are natural rings.

§1.3 On existence.

In the definition of §1.2, the phrase "one can find" has been used instead of "there exists" to emphasize the algorithmic foundation of the theory. However, the entire theory is constructed in this book according to Kroneckerian principles, which means that "one can find" and "there exists" are to be regarded as *synonymous*. For example, because Theorem 2

(§1.20) asserts that there exists a divisor with certain properties, the proof must provide an algorithm (with a simple *a priori* bound on the length of the computation) for *constructing* such a divisor. This meaning of "there exists" is the only one Kronecker recognized. Accordingly, the theorems of the theory say a great deal more than they appear to say to a reader who regards "there exists" as meaning "if one could prove there did not exist, then one could derive a contradiction."

§1.4 Preliminaries.

The development of the theory of divisors will require the following elementary consequences of the definition of natural rings:

Any finite set $a_1, a_2, \ldots, a_\nu$ of nonzero elements of a natural ring r, has a greatest common divisor c; that is, there is a $c \in r$ which divides all the a_i and is divisible by any $d \in r$ which divides all the a_i. This is true for $\nu = 2$ by axiom (2). Suppose it is true for $\nu - 1$, and let b be a g. c. d. of $a_1, a_2, \ldots, a_{\nu-1}$. A greatest common divisor c of b and a_ν divides all the a's, and any common divisor d of all the a's divides both b and a_ν and therefore divides c. Therefore, c is a g. c. d. of the a's, and the proposition follows by induction.

If a, b, and c are elements of r such that $c|ab$ but $c \nmid a$ and $c \nmid b$, then $c = c_1 c_2$, where neither c_1 nor c_2 is a unit. To prove this, let d be a g. c. d. of ab and cb. Then $b|d$, say $d = bc_1$, and the quotient c_1 divides both a and c. Set $c_2 = c/c_1$. Since c divides both ab and cb, $c|d$, that is, $c_1 c_2 | bc_1$. Therefore, $c_2 | b$. If $c_1 | 1$ then $c = c_1 c_2 | c_2$ gives $c|b$, contrary to assumption. If $c_2 | 1$ then $c | c_1$, which implies $c|a$, contrary to assumption. Therefore, neither c_1 nor c_2 is a unit.

Let f and g be polynomials with coefficients in r, and let a and b be greatest common divisors of the coefficients of f and g respectively. Then ab is a greatest common divisor of the coefficients of fg. The coefficients of fg have a g. c. d., call it c. Since $fg = ab(f/a)(g/b)$ and since $(f/a)(g/b)$ has coefficients in r, $ab|c$, say $c = abd$. Assume d is not a unit. Let an order be chosen for the indeterminates in f and g, and let the terms of polynomials be ordered lexicographically, that is, with one term preceding another if it contains the first indeterminate to a higher power, or, if they contain the first indeterminate to the same power, it contains the second indeterminate to a higher power, etc. Since ad does not divide a, at least one coefficient of f is not divisible by ad. Therefore, f can be written in the form $f = adf_1 + af_2$, where the coefficients of f_1 and f_2 are in r, where $f_2 \neq 0$, and where no coefficient of f_2 is divisible by d. Similarly, let $g = bdg_1 + bg_2$, where $g_2 \neq 0$ and no coefficient of g_2 is divisible by d. Since $fg = abd(df_1g_1 + f_2g_1 + f_1g_2) + abf_2g_2$, and all coefficients of fg are divisible by abd, all coefficients of f_2g_2 are divisible by d. The leading term of f_2g_2 is the product of the leading terms of f_2 and g_2. Therefore, the coefficient of the leading term of f_2g_2 is on the one hand divisible by d and on the other is a product of two elements of r, neither of which is divisible by d. As was shown above, it follows that $d = d_1d_2$, where neither d_1 nor d_2 is a unit. But then d_1 and d_2 have the same properties as d: they are not units, and all coefficients of fg are divisible by abd_i. Therefore, by the same argument, $d_1 = d_{11}d_{12}$ and $d_2 = d_{21}d_{22}$, where d_{ij} is not a unit and abd_{ij} divides all coefficients of fg. Therefore, $d = d_1d_2 = d_{11}d_{12}d_{21}d_{22} = \cdots$ can be written as a product of arbitrarily many factors, none of them units, contrary to axiom (3). Therefore, d must be a unit, say $1 = de$, which implies $c|ce = abde = ab$ and shows that ab is a g.c.d. of the coefficients of fg.

A polynomial f (in any number of indeterminates) with coefficients in a natural ring r is said to be *primitive* if it is not 0 and if 1 is a greatest common divisor of its coefficients. (Here, and always, the "coefficients" of f are the coefficients *after like terms have been combined*. For example, $2x + 3x$ is not a primitive polynomial with coefficients in $\mathbf{Z}$, but $2x + 3x^2$ and $2x + 3y$ are.)

By the above theorem, *a product of primitive polynomials is primitive.*

Let f be a nonzero polynomial with coefficients in r, and let c be a greatest common divisor of the coefficients of f. Then f/c *is primitive.* Because, if d is a g.c.d. of the coefficients of f/c, the above theorem shows cd is a g.c.d. of the coefficients of f; therefore, $cd|c$, which implies $d|1$, so any common divisor of the coefficients of f/c divides 1.

If 1 *is a greatest common divisor of* a_1, a_2, ..., a_ν, *then it is also a greatest common divisor of* a_1^j, a_2^j, ..., a_ν^j *for any integer* $j \geq 1$. To prove this, it will suffice to prove that 1 is a g.c.d. of a_1^j, a_2, a_3, ..., a_ν, because then each a can be raised to the power j in turn. Let $\pi = \sum a_i u_i$ where the u_i are indeterminates. Since π is primitive by assumption, π^j is primitive. Since the coefficients of π^j are a_1^j and elements of r divisible by at least one of a_2, a_3, ..., a_ν, any common divisor of a_1^j, a_2, a_3, ..., a_ν divides all coefficients of π^j and therefore must be a unit, as was to be shown.

GAUSS'S LEMMA. *Let k be the field of quotients of r. If the product of two monic polynomials in one indeterminate with coefficients in k has coefficients in r, then both factors must have coefficients in r.*

Otherwise stated, if $(a_0 x^m + a_1 x^{m-1} + \cdots + a_m)(b_0 x^n + b_1 x^{n-1} + \cdots + b_n)$, where the a's and b's are in r and x is an indeterminate, has all its coefficients divisible by $a_0 b_0$, then a_0 divides all the a's and b_0 divides all the b's. This follows, of course, from the theorem of Part 0 (an element of k is integral over r if and only if it is in r). It can also be deduced from the more elementary theorem above as follows.

If c and d are g.c.d.'s of the a's and b's respectively, then cd is a g.c.d. of the coefficients of the product $(\sum a_i x^{n-i})(\sum b_i x^{n-i})$. By assumption, $a_0 b_0$ divides all these coefficients. Thus $a_0 b_0$ divides cd, say $cd = e a_0 b_0$. Since $d|b_i$ for all i, and $c|a_0$, $cd|a_0 b_i$, $e a_0 b_0 | a_0 b_i$, $e b_0 | b_i$, $b_0 | b_i$. In the same way, $a_0 | a_i$, as was to be shown.

Gauss's lemma for polynomials in several indeterminates can be deduced from the case of one indeterminate using the technique of the deduction of Corollary 3, Part 0. (In the notation there, let X_ν be the indeterminate in which the polynomials f and g are monic, so that $\phi(f)$ and $\phi(g)$ are monic in T.) Thus, *if the product of two polynomials in several indeterminates with coefficients in k has coefficients in r, and if the polynomials are monic in one of the indeterminates, then both factors must have coefficients in r.*

§1.5 The basic theory.

The main objective of the theory is to define a *greatest common divisor* of a finite set of elements $\alpha_1, \alpha_2, \ldots, \alpha_\nu$, each of which is *algebraic* over r. An *algebraic extension* of r is a field containing r, all of whose elements are algebraic over r. It is natural to assume that the α's are contained in a fixed algebraic extension K of r.

The notion of greatest common divisor must be preceded by a notion of divisibility, which must in turn be preceded by a notion of integrality: an element divides another if the quotient is integral. An element α algebraic over a natural ring r is said to be *integral* over r if some power of it can be written as a linear combination of lower powers with coefficients in r. As can easily be shown (see Part 0), the elements of K integral over r form a *ring* in K.

The greatest common divisor of a finite subset $\alpha_1, \alpha_2, \ldots, \alpha_\nu$ of K will be defined by defining a divisor which depends on *more* than the finite subset—specifically, it will depend on a *polynomial* whose *coefficients* are the α's—and then showing that the divisor so defined has all the desired properties. Among the desired properties is that the divisor represented by a polynomial with coefficients in K depends only on the coefficients of the polynomial. (See Corollary (11) of §1.12.)

§1.6 Definitions.

Let K be an algebraic extension of r. A polynomial f, in any number of indeterminates and with coefficients in K, is said to *divide* another such polynomial g if $g = fq$, where q is a polynomial whose coefficients are elements of K *integral* over r.

We will say that a polynomial f (in any number of indeterminates and with coefficients in K) *represents a divisor*, and will denote the divisor it represents by $[f]$. If f and g are such polynomials, we will say that *the divisor represented by f divides g*—in symbols $[f]|g$—if there is a primitive polynomial π with coefficients in r such that f divides $g\pi$. In other words, $[f]|g$ means that $g\pi = fq$, where π is a primitive polynomial with coefficients in r and q is a polynomial with coefficients integral over r.

The divisor represented by f_1 will be said to *divide* the divisor represented by f_2—in symbols, $[f_1]|[f_2]$—if $[f_2]|g$ implies $[f_1]|g$.

Two polynomials f_1 and f_2 with coefficients in K will be said to *represent the same divisor*—in symbols, $[f_1] = [f_2]$—if $[f_1]|[f_2]$ and $[f_2]|[f_1]$.

§1.7.

The contemporary style of mathematics trains mathematicians to ask "What is a divisor?" and to want the answer to be framed in terms of set theory. Those trained in this tradition will want to think of a divisor as *an equivalence class of polynomials*, when equivalence of polynomials is the property of representing the same divisor. I believe, however, that instead of asking what a divisor *is* one should ask what it *does*. It *divides* things. Specifically, it divides (or does not divide) polynomials with coefficients in K. The definition of what a divisor does involves a given polynomial, and two polynomials represent the same divisor if the corresponding divisors do the same thing.

§1.8 First Propositions.

PROPOSITIONS. (1) If $[f]|g_1$ and $[f]|g_2$, then $[f]|(g_1 \pm g_2)$.

(2) If $[f]|\beta$ for all coefficients β of g, then $[f]|g$.

(3) If $[f]|g$ and $[g]|h$, then $[f]|h$.

(4) $[f]|g$ if and only if $[f]|[g]$.

(5) If f has coefficients in r, if $f \neq 0$, and if d is a greatest common divisor of the coefficients of f, then $[f] = [d]$.

(6) If $[f_1]|[f_2]$, then $[f_1 f_3]|[f_2 f_3]$ for all polynomials f_3 with coefficients in K.

PROOFS: (1) If $fq_1 = g_1\pi_1$ and $fq_2 = g_2\pi_2$, then $f(q_1\pi_2 \pm q_2\pi_1) = (g_1 \pm g_2)\pi_1\pi_2$. The coefficients of $q_1\pi_2 \pm q_2\pi_1$ are integral over r because the coefficients of q_1, q_2, π_1, π_2 are, and because the elements of K integral over r form a ring. The product $\pi_1\pi_2$ of two primitive polynomials with coefficients in r is itself primitive, as was proved above. Therefore, (1) follows. (2) $[f]|\beta$ implies $fq = \beta\pi$ for π primitive and for q with coefficients integral over r. If this equation is multiplied by any monomial with coefficient 1, it shows that any monomial with coefficient β is divisible by $[f]$. Therefore, (2) follows from (1). (3) If $[f]|g$, say $fq_1 = g\pi_1$, and if $[g]|h$, say $gq_2 = h\pi_2$, then $fq_1q_2 = g\pi_1q_2 = h\pi_1\pi_2$, which shows $[f]|h$. (4) Since $[g]|g$ (set $q = \pi = 1$), $[f]|[g]$ implies $[f]|g$. Conversely, if $[f]|g$ and if $[g]|h$ then $[f]|h$ by (3); thus $[f]|[g]$. (5) f/d is primitive (§1.4). Therefore, $[f]|d$ $(q = 1, \pi = f/d)$ and $[d]|f$ $(q = f/d, \pi = 1)$, so $[f] = [d]$ by (4). (6) If $f_1q = f_2\pi$, then $f_1f_3q = f_2f_3\pi$ for all f_3, and (6) follows.

§1.9.

PROPOSITION (7). If $\alpha, \beta \in K$, and $\alpha \neq 0$, then $[\alpha]|\beta$ if and only if β/α is integral over r.

PROOF: If β/α is integral over r, then $\alpha \cdot (\beta/\alpha) = \beta \cdot 1$ shows that $[\alpha]|\beta$. Conversely, suppose $[\alpha]|\beta$, that is, $\alpha q = \beta\pi$. With

$\gamma = \beta/\alpha$, this is simply the statement that there is a primitive polynomial π such that $\gamma\pi$ has coefficients integral over r. It is to be shown that γ is then integral over r.

Since γ is algebraic over r, $h(\gamma) = 0$ for some nonzero polynomial $h(x)$ in one indeterminate with coefficients in r. Division of $h(x)$ by a g. c. d. of its coefficients gives a primitive $h(x)$ for which $h(\gamma) = 0$. Therefore, one can assume without loss of generality that $h(x)$ is primitive. If $h(x)$ is not irreducible, then, by axiom (4), $h(x) = h_1(x)h_2(x)$, where h_1 and h_2 have coefficients in r. Since h is primitive, neither h_1 nor h_2 has degree 0, so both have degree less than $\deg h$. Since either $h_1(\gamma) = 0$ or $h_2(\gamma) = 0$, and both h_1 and h_2 are primitive, it follows that, if h is not irreducible, there is a polynomial of smaller degree than h with the same properties as h. Therefore, by the principle of infinite descent, one can assume without loss of generality that $h(x)$ is a polynomial *irreducible* over r for which $h(\gamma) = 0$, say $h(x) = b_0 x^n + b_1 x^{n-1} + \cdots + b_n$.

Suppose $a \in r$ has the property that $a\gamma$ is integral over r, say $\phi(a\gamma) = 0$ where ϕ is monic with coefficients in r. The Euclidean algorithm applied to $\phi(ax)$ and $h(x)$, regarded as polynomials with coefficients in the field of quotients k of r, gives a polynomial $d(x)$ of the form $d(x) = s(x)\phi(ax) + t(x)h(x)$ which divides both $\phi(ax)$ and $h(x)$ as polynomials with coefficients in k. Clearing denominators then gives equations of the form $c_1 d(x) = s_1(x)\phi(ax) + t_1(x)h(x)$, $c_2\phi(ax) = q_1(x)d(x)$, and $c_3 h(x) = q_2(x)d(x)$, where c_1, c_2, and c_3 are nonzero elements of r and $d(x)$, $s_1(x)$, $t_1(x)$, $q_1(x)$, and $q_2(x)$ have coefficients in r. There is no loss of generality in assuming that $d(x)$ is primitive (any common factor of its coefficients could be absorbed into c_1, $q_1(x)$, and $q_2(x)$). Then, since $h(x)$ is primitive (it is irreducible), c_3 divides all coefficients of $q_2(x)$ (1 times a g. c. d. of the coefficients of $q_2(x)$ is a g. c. d. of the coefficients of $c_3 h(x)$). Therefore, $h(x) = q_3(x)d(x)$, where $q_3(x)$ has coefficients in r. Since $h(x)$ is irreducible over r, either $q_3(x)$ or $d(x)$ must be a unit of r. Since $c_1 d(\gamma) =$

$s_1(\gamma)\phi(a\gamma) + t_1(\gamma)h(\gamma) = s_1(\gamma) \cdot 0 + t_1(\gamma) \cdot 0 = 0$, $d(\gamma) = 0$, so $\deg d > 0$. Therefore $g_3(x)$ is a unit of r, say $q_3(x) = e \in r$. Then $h(x) = ed(x)$, $ec_2\phi(ax) = q_1(x)h(x)$. Since ec_2 divides all coefficients of $q_1(x)h(x)$ and $h(x)$ is primitive, ec_2 divides all coefficients of $q_1(x)$, and $\phi(ax) = q_4(x)h(x)$, where $q_4(x)$ has coefficients in r. Thus $\phi(y) = q_4(y/a)h(y/a)$. Let m be the degree of q_4. Then $a^{m+n}\phi(y) = (a^m q_4(y/a))(a^n h(y/a))$. The factors on the right have coefficients in r, and their product $a^{m+n}\phi(y)$ has all its coefficients divisible by its leading coefficient a^{m+n}. Therefore, by Gauss's Lemma, all coefficients of $a^n h(y/a) = b_0 y^n + ab_1 y^{n-1} + a^2 b_2 y^{n-2} + \cdots + a^n b_n$ are divisible by its leading coefficient b_0. Therefore, b_0 divides all coefficients of $a^n h(x)$. Since $h(x)$ is primitive, b_0 divides a^n.

Thus, if γa is integral over r for all coefficients a of π, b_0 divides a^n for all coefficients a of π. Since 1 is a g.c.d. of the a's, it is a g.c.d. of their nth powers (§1.4), and b_0 divides 1, say $b_0 e_0 = 1$. Therefore γ is a root of the monic polynomial $e_0 h(x)$ with coefficients in r, which means γ is integral over r, as was to be shown.

§1.10.

LEMMA (8). *If f is a nonzero polynomial with coefficients in K, then $[f]\|[c]$ for some nonzero $c \in r$.*

PROOF: Adjunction of the coefficients of f to the field of quotients k of r gives a subfield K' of K of finite degree over k. Let $\omega_1, \omega_2, \ldots, \omega_\mu$ span K' over k. Each $f\omega_i$ for $i = 1, 2, \ldots, \mu$ can then be written in the form $\sum_{j=1}^{\mu} m_{ij}\omega_j$, where the m_{ij} are polynomials in the same indeterminates as f with coefficients in k. Let X be a new indeterminate, and let $F(X) = \det(XI - M)$, where I is the $\mu \times \mu$ identity matrix and M is the $\mu \times \mu$ matrix whose entries are the m_{ij}. Then $F(X) = X^\mu + F_1 X^{\mu-1} + \cdots + F_\mu$, where the F_i are polynomials in the same indeterminates as f with coefficients in k. Substitution of f for X in $XI - M$ gives a $\mu \times \mu$ matrix $fI - M$ with entries from the ring of polynomials in the indeterminates of

f with coefficients in K. In this ring, the $\mu \times \mu$ system of homogeneous linear equations $(fI - M)Y = 0$ has the non-trivial solution $Y = (\omega)$, where (ω) denotes the column matrix whose entries are $\omega_1, \omega_2, \ldots, \omega_\mu$. Therefore, $\det(fI - M) = 0$, that is, $f^\mu + F_1 f^{\mu-1} + \cdots + F_\mu = 0$. Thus, $F_\mu = fg$, where F_μ is a polynomial with coefficients in k and $g = -f^{\mu-1} - F_1 f^{\mu-2} - \cdots - F_{\mu-1}$ is a polynomial with coefficients in K. Each coefficient β of g satisfies an equation of the form $b_0 \beta^n + b_1 \beta^{n-1} + \cdots + b_n = 0$, where the $b_i \in r$; such an equation implies that $b_0 \beta$ is integral over r, because multiplication by b_0^{n-1} gives $(b_0 \beta)^n + b_1 (b_0 \beta)^{n-1} + b_0 b_2 (b_0 \beta)^{n-2} + \cdots + b_0^{n-1} b_n = 0$. Thus, for each coefficient β of either g or F_μ, there is a $b \in r$ such that $b\beta$ is integral over r. Multiplication of $F_\mu = gf$ by the product of all these b's gives an equation of the form $h = qf$, where h has coefficients in r and q has coefficients in K integral over r. Thus, $[f]$ divides $[h]$. Since $[h] = [c]$ for $c \in r$ by (5), the Lemma follows.

§1.11 The Main Proposition.

PROPOSITION (9). $[f] | \alpha$ *for all coefficients α of f.*

This is the final proposition in the basic theory of divisors. The other propositions have all been deduced from elementary properties of natural rings and algebraic extensions of them. This last one depends, however, on a less elementary theorem. It can be deduced, for example, from Corollary 4 of the theorem of Part 0:

PROOF: If $f = 0$, there is nothing to prove. Otherwise, by (8), $fq = c\pi$ for some nonzero $c \in r$ and for some primitive polynomial π with coefficients in r. Thus, the product of the polynomials f and q/c has coefficients in r. If α is any coefficient of f and γ any coefficient of q, then, by Corollary 4, $\alpha\gamma/c$ is integral over r. Set $q_1 = \alpha q/c$. Then $fq_1 = f\alpha q/c = \alpha c\pi/c = \alpha\pi$, where q_1 has coefficients integral over r and π is primitive. Therefore, $[f] | \alpha$.

§1.12 Concluding corollaries.

COROLLARIES. (10) $[f]$ *is the greatest common divisor of the coefficients α of f in the sense that $[f]|\alpha$ for all α and $[f]$ is divisible by any divisor $[g]$ which divides all coefficients α of f.*

(11) *If f and g have the same coefficients, then $[f] = [g]$.*

(12) *Given polynomials f and g with coefficients in K, one can test whether $[f]|g$ as follows. If $f = 0$, then $[f]|g$ if and only if $g = 0$. Otherwise, the construction of Lemma (8) gives an equation $fq = c\pi$, where $c \in r$, $c \neq 0$, π is primitive (coefficients in r) and q has coefficients in K integral over r. Then $[f]|g$ if and only if the coefficients of gq/c are integral over r.*

(13) $[1]|[g]$ *if and only if the coefficients of g are integral over r.*

DEDUCTIONS: (10) follows from (9), (2), and (4). (11) follows from (10). If, in (12), all coefficients of $q_1 = gq/c$ are integral over r, then the equation $fq_1 = fqg/c = c\pi g/c = g\pi$ shows $[f]|g$ directly from the definition. Conversely, if $[f]|g$, then, by (4) and (6), $[fq]|[gq]$. But $[fq] = [c\pi] = [c]$ by (5). Thus $[c]|[gq]$. By (9), it follows that $[c]$ divides all coefficients of gq, hence, by (7), that all coefficients of gq/c are integral over r. Finally, (13) is the case $f = c = \pi = q = 1$ of (12).

§1.13 Testing for divisibility.

Application of the divisibility test (12) requires that one be able to determine, for a given $\alpha \in K$, whether α is integral over r. As was shown in §1.9, there is an irreducible $h(x)$ with coefficients in r such that $h(\alpha) = 0$; if $a\alpha$ is integral over r, then the leading coefficient of h divides a^n, where $n = \deg h$. In particular, if α itself is integral over r, then the leading coefficient of h is a unit of r. Thus, to determine whether α is integral over r, it suffices to find a relation of the form $h(\alpha) = 0$, where $h(x)$ is irreducible over r; α is integral over r if and only if the leading coefficient of h is a unit of r.

§1.14 The Group of Nonzero Divisors.

Proposition (6) of §1.8 shows that $[f_1] = [f_2]$ implies $[f_1 g] = [f_2 g]$ for all polynomials g with coefficients in K. Therefore, it is valid to define the *product* of the divisors $[f]$ and $[g]$ to be $[fg]$. The product operation defined in this way is obviously commutative and associative. Moreover, $[1]$ is an identity element with respect to it. Finally, if $[f] \neq [0]$, then, by Lemma (8), $fq = c\pi$ for some $c \in r$, for some primitive polynomial π with coefficients in r, and for some polynomial q with coefficients in K integral over r, which shows that $[f][q/c] = [\pi] = [1]$ (by (5)); in particular, every divisor $[f] \neq [0]$ has a *multiplicative inverse*. In short, nonzero divisors with the operation of multiplication just defined form a *commutative group*.

An *integral* divisor $[g]$ is a divisor represented by a polynomial g, all of whose coefficients are integral over r. By Corollary (13), $[g]$ is integral if and only if $[1]|[g]$; therefore, if f is any polynomial which represents an integral divisor, all coefficients of f are integral over r. Because elements of K integral over r form a ring, a product of integral divisors is integral. Moreover, since it was shown above that every nonzero $[f]$ has a multiplicative inverse $[q/c] = [q]/[c]$, every nonzero $[f]$ can be written in the form $[q]^{-1}[c]$ in the group of nonzero divisors. Thus *every nonzero divisor can be written as a quotient of integral divisors* in the group of nonzero divisors.

Finally, if $[f]$ and $[g]$ are nonzero divisors, then $[f]|[g]$ *in the sense defined in* §1.6 *if and only if* $[f]^{-1}[g]$ *is an integral divisor.* For, if $g\pi = fq$, then $[f]^{-1}[g] = [f]^{-1}[g\pi] = [f]^{-1}[fq] = [f]^{-1}[f][q] = [q]$, which is integral, and, conversely, if $[f]^{-1}[g] = [q]$ is integral, then $[g] = [fq]$ is divisible by $[f]$ because $[f]|[fq]$ follows directly from the definition.

§1.15 Greatest common divisors.

PROPOSITION (14). *Any finite subset* $[f_1], [f_2], \ldots, [f_\nu]$ *of the group of nonzero divisors has a greatest common divisor* $[g]$.

*That is, there is a nonzero divisor $[g]$ such that $[g]\|[f_i]$ for $i = 1$,
$2, \ldots, \nu$, and such that if $[h]$ satisfies $[h]\|[f_i]$ for $i = 1, 2, \ldots, \nu$
then $[h]\|[g]$. Of course, by the very definition, any two greatest
common divisors of $[f_1], [f_2], \ldots, [f_\nu]$ are equal.*

PROOF: By Corollary (11), $[f_i]$ is not changed if all indeterminates in the polynomial f_i are changed. Therefore, there is no
loss of generality in assuming that no indeterminate of f_i occurs in any other f_j. Then the coefficients of $f_1 + f_2 + \cdots + f_\nu$ are
the union of all coefficients of the f_i. Let $g = f_1 + f_2 + \cdots + f_\nu$.
Then $[g]\|[f_i]$ for all i, and $[h]\|[f_i]$ for all i implies $[h]\|[g]$, both
by Corollary (10). Therefore $[g]$ is the desired g.c.d.

Notation: The greatest common divisor of $[f_1], [f_2], \ldots, [f_\nu]$
will be denoted $[f_1, f_2, \ldots, f_\nu]$. In particular, if $\alpha_1, \alpha_2, \ldots,$
α_μ are the coefficients of f, then $[f] = [\alpha_1, \alpha_2, \ldots, \alpha_\mu]$.

PROPOSITIONS. (15) $[f, g, h]$ is the greatest common divisor of
$[f]$ and $[g, h]$.
  (16) $[f][g, h] = [fg, fh]$.
  (17) If $[f, g] = [1]$ and $[h]$ is integral, then $[f, gh] = [f, h]$.
  (18) If $[f, g] = [f, h] = [1]$, then $[f, gh] = [1]$.
  (19) If $[f, g] = [1]$ and $[h]$ is integral, then $[f, h][g, h] =$
$[fg, h]$.

PROOFS: (15) A divisor which divides $[f]$, $[g]$, and $[h]$ divides $[f]$ and $[g, h]$, and conversely. (16) As in the proof of
(14), change the indeterminates of f, g, and h—which does
not change any of the divisors $[f]$, $[g, h]$, $[fg]$, $[fh]$—so that
no indeterminate occurs in more than one of them. Then
$[g + h] = [g, h]$ and $[fg, fh] = [fg + fh] = [f][g + h] = [f][g, h]$.
(17) By (16), if $[f, g] = [1]$, then $[fh, gh] = [1][h] = [h]$. Then
$[f, h] = [f, fh, gh]$ by (15), and, since $[f] = [f][1] = [f][1, h] =$
$[f, fh]$, it follows that $[f, fh, gh] = [f, gh]$, also by (15). (18)
is a special case of (17). (19) Let f, g, and h involve distinct indeterminates. Then $[f, h][g, h] = [f + h][g + h] =$
$[fg + fh + hg + h^2] = [fg, fh, gh, h^2]$. Since $[f, g, h]$ di-

vides $[f,g] = [1]$ and f and g are integral, $[f, g, h] = [1]$, $[fh,$ $gh, h^2] = [h]$ by (16) and $[f, h][g, h] = [fg, h]$ by (15).

§1.16 Ambient fields.

Up to this point, the field K has remained entirely in the background. Kronecker deliberately structured his theory (see the beginning of §17 of [**Kr1**]) in such a way that K plays no role. Division of polynomials is a *rational* operation, in that if $F = GH$, where F, G, and H are polynomials, in any number of indeterminates, with coefficients in a field K, then the coefficients of H are determined as rational functions of the coefficients of F and G. (For polynomials in one indeterminate, this follows from the division algorithm. The proof for polynomials in several indeterminates is then a simple induction.) Therefore, if f and g are polynomials with coefficients in K and if π is a primitive polynomial with coefficients in r such that $f|g\pi$—that is, $g\pi = fq$, where q has coefficients integral over r—then the same is true (with the same π and q) when K is replaced by any larger or smaller field K', provided K' contains all coefficients of f and g. Therefore, any statement about divisors represented by polynomials with coefficients in K is true for divisors represented by polynomials with coefficients in any algebraic extension K' of r such that the divisors involved can all be represented by polynomials with coefficients in K'.

The above formulation of the theory is even more independent of the ambient field than was Kronecker's, because Kronecker used *norms* in defining the basic concept $[f]|g$. (Specifically, for a linear polynomial f with coefficients *integral* over r, he sets [**Kr1**, §14] $Nf = P \cdot \mathrm{Fm}(f)$, where $P \in r$ and where $\mathrm{Fm}(f)$ is a primitive polynomial with coefficients in r, and defines $[f]|g$ to mean that $(f/\mathrm{Fm}(f))|g$; since $f/\mathrm{Fm}(f) = Pf/Nf$, $[f]|g$ then means that P divides $g \cdot (Nf/f)$, which is equivalent to the meaning given to $[f]|g$ in §1.6, as Corollary (12) shows.)

Norms are defined in the next section in order to have them available for the proofs of Theorems 1 and 2 below. The norm of a polynomial with coefficients in a field K depends on the ambient field (albeit in a simple way—passing to a larger field raises the norm to a power), but Theorems 1 and 2 are independent of the ambient field that is used, provided only that it contains all the divisors in question.

§1.17 Norms.

An algebraic extension K of a natural ring r will be said to be of *finite degree* if its dimension as a vector space over the field of quotients k of r is finite.

DEFINITION. Let f be a polynomial with coefficients in an algebraic extension K of r of finite degree, and let $\alpha_1, \alpha_2, \ldots,$ α_λ be a basis of K over k, the field of quotients of r. Then the equations $f \cdot \alpha_i = \sum_{j=1}^{\lambda} \phi_{ij}\alpha_j$ define polynomials ϕ_{ij} in the same indeterminates as f with coefficients in k. The *norm of f relative to the extension $K \supset r$*, denoted $N_K f$, is the determinant of the $\lambda \times \lambda$ matrix whose entries are ϕ_{ij}. This definition of $N_K f$ is independent of the choice of the basis of K over k, because any other basis is transformed to the chosen one by multiplication by an invertible matrix M with entries in k, which changes the matrix with entries ϕ_{ij} by multiplying it on one side by M and on the other by M^{-1}, an operation which leaves its determinant unchanged. Note that $N_K f$ is a polynomial, in the same indeterminates as f, with coefficients in k. When the extension $K \supset r$ under consideration does not need emphasis, $N_K f$ will be denoted simply Nf.

PROPOSITION. *If the coefficients of f are integral over r, then $Nf = fq$, where q can be expressed in the form*

$$q = \pm(f^\nu + B_1 f^{\nu-1} + \cdots + B_\nu)$$

with ν a positive integer and $B_1, B_2, \ldots, B_\nu$ polynomials in

the same indeterminates as f with coefficients in r. In particular, q has coefficients integral over r.

PROOF: There exist elements ω_1, ω_2, ..., ω_μ of K such that
any coefficient of f times any ω is a linear combination of
ω's with coefficients in r (Proposition 1 of Part 0). Thus,
$f\omega_i = \sum m_{ij}\omega_j$, where the m_{ij} are polynomials with coefficients in r, and the argument of §1.10 gives a polynomial
$F(X) = X^\mu + F_1 X^{\mu-1} + \cdots + F_\mu$, with coefficients in r,
for which $F(f) = 0$. By the Remainder Theorem, $F(X) =$
$(X - f)Q(X)$, where $Q(X)$ is a polynomial with coefficients in
K. The norm of $F(X)$ is the determinant of $F(X) \cdot I$, where I is
the $\lambda \times \lambda$ identity matrix, which is $F(X)^\lambda$. On the other hand,
the norm of a product is the product of the norms (because
the determinant of a product is the product of the determinants), so $F(X)^\lambda = N(X - f)NQ(X)$. By Gauss's Lemma,
since $F(X)^\lambda$ is monic with coefficients in r and $N(X - f)$ is
monic with coefficients in k, $N(X - f)$ has coefficients in r, say
$N(X-f) = X^\lambda + A_1 X^{\lambda-1} + \cdots + A_\lambda$, where the A_i are polynomials in the indeterminates of f with coefficients in r. Substitution of $X = 0$ in $N(X - f)$ gives $(-1)^\lambda Nf = A_\lambda$. Substitution of $X = f$ in the matrix of which $N(X - f)$ is the determinant gives a matrix M which satisfies $M\alpha = 0$, where α is the
column matrix giving the chosen basis of K over k, and where
0 is the column matrix of λ zeros. Thus, $\det M = 0$, which is to
say, $f^\lambda + A_1 f^{\lambda-1} + \cdots + A_\lambda = 0$, that is, $Nf = (-1)^\lambda A_\lambda = fq$,
where $q = (-1)^{\lambda-1}(f^{\lambda-1} + A_1 f^{\lambda-2} + \cdots + A_{\lambda-1})$, as was to
be shown.

COROLLARIES. (1) If $[f]$ is integral, then $[f]\,|\,[Nf]$. In partic-
ular, $[Nf]$ is integral.
 (2) If $[f]\,|\,[g]$, then $[Nf]\,|\,[Ng]$.
 (3) $[Nf]$ depends only on $[f]$ and can therefore be defined
to be $N[f]$, the norm of the divisor.
 (4) If $N[f] = [0]$, then $[f] = [0]$.
 (5) Given a nonzero integral divisor $[f]$ in K, there is an

integer ν such that any representation of $[f]$ as a product of more than ν integral divisors in K contains at least one factor equal to $[1]$.

DEDUCTIONS: (1) is immediate. (2) If $fq = g\pi$, then $Nf \cdot Nq = Ng \cdot \pi^\lambda$. Since $[Nq]$ is integral by (1) and π^λ is primitive, $[Nf]|[Ng]$. (3) follows immediately from (2). (4) If $[f] \neq [0]$, then $[f] = [g]$, where g is a polynomial with a nonzero constant term, call it g_0. Directly from the definition, the constant term of Ng is the determinant of the matrix which represents multiplication by g_0 relative to a basis of K over k. Since multiplication by g_0 is invertible, so is the matrix which represents it; therefore, the constant term of Ng is nonzero, and $N[f] = [Ng] \neq [0]$, as was to be shown. (5) The norm of a factorization of $[f]$ is a factorization of $[Nf]$. Since $[Nf]$ and all its factors obtained in this way are of the form $[c]$ for $c \in r$, Axiom (3) of the definition of a natural ring implies that, for ν sufficiently large, at least one of the factors must divide $[1]$. By (1), an integral divisor whose norm divides $[1]$ is $[1]$.

§1.18 Factorization of divisors.

Let K be a fixed algebraic extension of the natural ring r. An integral divisor in K is *reducible in K* if it can be written as a product of two integral divisors in K, neither of which is $[1]$; otherwise it is *irreducible in K*. An integral divisor in K is *prime in K* if it is not $[1]$ and if it divides a product of integral divisors only when it divides one of the factors; otherwise stated, an integral divisor $[f]$ in K is prime if the integral divisors in K *not* divisible by $[f]$ form a nonempty set closed under multiplication.

PROPOSITION. *An integral divisor other than $[1]$ is irreducible in K if and only if it is prime in K.*

PROOF: If $[f]$ is prime in K and if $[f] = [g][h]$, where $[g]$ and $[h]$ are integral divisors in K, then $[f]|[g][h]$, which implies $[f]|[g]$

or $[f]|[h]$. If $[f]|[g]$, say $[g] = [f][q]$, where $[q]$ is integral, then $[f] = [g][h] = [f][q][h]$, $[1] = [q][h]$, $[h]|[1]$, $[h] = [1]$. Similarly, if $[f]|[h]$, then $[g] = [1]$, so $[g]$ or $[h]$ is $[1]$, as was to be shown.

Suppose now that $[f] \neq [1]$ is an integral divisor irreducible in K, and suppose that $[f]|[g][h]$, where $[g]$ and $[h]$ are divisors in K. If $[f,g] = [1]$ then $[f] = [f,gh] = [f,h]$ by Proposition (17) of §1.15, that is, $[f]|[h]$. If $[f,g] \neq [1]$, then, since $[f] = [f,g][q]$ for some integral divisor $[q]$, and $[f]$ is irreducible, $[q]$ must be $[1]$, $[f]$ must be $[f,g]$, and $[f]$ must divide $[g]$. Thus, $[f]$ is prime.

To factor an integral divisor in K into divisors *irreducible* in K is therefore the same as to factor it into divisors *prime* in K. When $K \supset r$ is an extension of finite degree, Corollary (5) of §1.17 guarantees, in a weak sense, that every integral divisor in K can be expressed as a product of divisors prime in K: Given any expression of $[f] \neq [1]$ as a product of integral divisors $[f] = [g_1][g_2] \cdots [g_\mu]$, none of which is equal to $[1]$, either the factors $[g_i]$ are all prime or at least one of them is not prime, that is, at least one of them is not irreducible, and the factorization can be carried further. The corollary gives an upper bound on the number of factors in such a factorization, and shows that this factorization process—beginning with the trivial factorization $[f] = [f]$—must terminate with an expression of $[f]$ as a product of divisors prime in K.

These prime factorizations of integral divisors in K "exist" in a very weak sense, however, because the process is not constructive. It lacks an algorithm which, for a given integral divisor $[f]$ in K, *either* expresses $[f]$ as a product of two integral divisors in K, neither of them equal to $[1]$, *or* proves that $[f]$ is irreducible in K. Such an algorithm is given in §2.1 in the special case $r = \mathbf{Z}$. Kronecker very briefly indicates an algorithm [**Kr1**, §18, end of the second paragraph] in the cases $r = \mathbf{Z}[x_1, x_2, \ldots, x_n]$.

The existence of such an algorithm for general r appears to be a subtler question than the ones dealt with here. For this reason, the question of factoring arbitrary integral divisors into divisors prime (irreducible) in a given field is avoided in what follows. Although factorization into primes has generally been regarded since Dedekind's time as the cornerstone of the theory, the lack of such factorizations does not seem to be any impediment at all. In fact, it forces one to rethink the fundamentals of the theory and to base proofs on more elementary and more constructive arguments.

§1.19 Two Basic Theorems.

THEOREM 1. *Let a finite number of divisors* $[f_1]$, $[f_2]$, $\ldots$, $[f_\mu]$ *be given. There exist relatively prime integral divisors* $[g_1]$, $[g_2]$, $\ldots$, $[g_\nu]$ *such that each* $[f_i]$ *is a product of powers of* $[g]$'s; *that is,* $[g_i, g_j] = [1]$ *whenever* $i \neq j$, *and there exist integers* σ_{ij} *such that* $[f_i] = [g_1]^{\sigma_{i1}}[g_2]^{\sigma_{i2}} \cdots [g_\nu]^{\sigma_{i\nu}}$ *for* $i = 1$, $2, \ldots, \mu$.

PROOF: The $[f]$'s can be expressed as quotients of integral divisors, say $[f_i] = [f_i']/[f_i'']$. Since a set of $[g]$'s for the 2μ integral divisors $[f_i']$, $[f_i'']$ will serve as a set of $[g]$'s for the given $[f]$'s, there is no loss of generality in assuming that the given $[f]$'s are *integral*. Consider the integral divisor which is the product $[f_1][f_2] \cdots [f_\mu]$ of the $[f]$'s. If, for some pair of indices, $[f_i, f_j]$ is a *proper* divisor of $[f_i]$—that is, $[f_i, f_j]$ is neither $[1]$ nor $[f_i]$—then the factorization $[f_1][f_2] \cdots [f_\mu]$ can be *refined* by replacing $[f_i]$ with two factors $[f_i, f_j][h]$, where $[h] = [f_i][f_i, f_j]^{-1} \neq [1]$. To find a set of $[g]$'s for the factors of $[f_1][f_2] \cdots [f_\mu]$, it will suffice to find a set of $[g]$'s for the factors of the refined factorization. Let this construction be repeated. It will produce greater and greater refinements of the factorization $[f_1][f_2] \cdots [f_\mu]$ unless a refinement is reached in which any two factors are either relatively prime ($[f_i, f_j] = [1]$) or identical ($[f_i, f_j] = [f_i] = [f_j]$). Let $[g_1]$, $[g_2]$, $\ldots$, $[g_\nu]$ be the distinct factors in such a factorization. Then $[f_1][f_2] \cdots [f_\mu] =$

$[g_1]^{\tau_1}[g_2]^{\tau_2}\cdots[g_\nu]^{\tau_\nu}$ and the factorization on the right is a refinement of the one on the left. Therefore, each $[f_i]$ is a product of powers of the $[g]$'s, as desired, and the refinement process, if it terminates, results in a set of $[g]$'s.

Let $K \supset r$ be the extension generated by the coefficients of the polynomials $f_1, f_2, \ldots, f_\mu$. The above construction lies entirely within the field K, which is an extension of r of finite degree. Therefore, by Corollary (5) of §1.17, there is an upper bound on the number of factors other than $[1]$ in a factorization of $[f_1][f_2]\cdots[f_\mu]$. Therefore, the above process must terminate, and Theorem 1 is proved.

PROPOSITION. *Let* $[g_1]$, $[g_2]$, $\ldots$, $[g_\nu]$ *be relatively prime integral divisors. Then* $\prod[g_j]^{\sigma_j}$ *divides* $\prod[g_j]^{\tau_j}$, *for* σ_1, σ_2, $\ldots$, σ_ν, τ_1, τ_2, $\ldots$, $\tau_\nu \in \mathbf{Z}$ *only if* $\sigma_j \le \tau_j$ *for all* j.

PROOF: Because both divisors can be multiplied by high powers of $\prod[g_j]$, there is no loss of generality in assuming that the σ's and τ's are all positive. If $\sigma_i > \tau_i$ and $\prod[g_j]^{\sigma_j}$ divides $\prod[g_j]^{\tau_j}$, then $[g_i]$ divides $[g_i]^{\sigma_i-\tau_i}\prod_{j\neq i}[g_j]^{\sigma_j}$ which in turn divides $\prod_{j\neq i}[g_j]^{\tau_j}$, contrary to Proposition (18), §1.15.

COROLLARIES. (1) *Given two factorizations of an integral divisor into integral divisors, there is a third factorization into integral divisors which is a refinement of both.*

(2) *Two representations of a divisor as a product of powers of distinct divisors prime in a field* $K \supset r$ *are identical, except possibly for the order of the factors.*

(3) *An integral divisor in* K *which divides a product of divisors prime in* K *is equal to the product of a subset of those divisors.*

DEDUCTIONS: (1) Let $[f_1][f_2]\cdots[f_\mu] = [h_1][h_2]\cdots[h_m]$ be the given factorizations. By the theorem, there exist $[g_1]$, $[g_2]$, $\ldots$, $[g_\nu]$ such that each $[f_i]$ and each $[h_i]$ is a product of powers of $[g]$'s. The Proposition implies that these powers are all nonnegative. Thus, the product of the $[f]$'s is a product of powers of $[g]$'s, and this product of powers of the $[g]$'s is a refinement of both of the given factorizations. (2) By the theorem, there exist relatively prime integral divisors $[g_1]$, $[g_2]$, $\ldots$, $[g_\nu]$ such that the factors of both factorizations are products of powers of the $[g]$'s. But since these factors, call them $[f_i]$, are irreducible, the factorizations $[f_i] = \prod [g_j]^{\sigma_{ij}}$ must have $\sigma_{ij} \geq 0$ (by the Proposition) and must have one $\sigma_{ij} = 1$ and the rest $= 0$ (by irreducibility). In short, the $[f_i]$ are *included* among the $[g_i]$. Both factorizations are therefore of the form $\prod [g_j]^{\sigma_j}$, and the uniqueness of such a factorization follows from the Proposition. (3) A common refinement of $[f][q] = [h_1][h_2] \cdots [h_\nu]$, where $[f]$ and $[q]$ are integral and the $[h_i]$ are prime in K, gives, in particular, $[f]$ as a product of integral divisors, each of which divides one of the $[h_i]$. Since the $[h_i]$ are irreducible, this gives, when factors $[1]$ are ignored, $[f]$ as a product of $[h]$'s. The exponent of any one $[h]$ in $[f]$ is of course at most its exponent in $[f][q]$.

§1.20.

THEOREM 2. *Let $[f]$ and $[g]$ be nonzero integral divisors. Then there is an integral divisor $[h]$ such that $[g, h] = [1]$ and $[f][h] = [\alpha]$ for some $\alpha \in K$.*

Otherwise stated, $[f]$ divides a "principal" divisor $[\alpha]$ with a quotient $[h]$ relatively prime to $[g]$. (Cf. Gauss, D. A., Art. 228.)

PROOF: Let $[g]$ be written as a product $[g] = [g_1][g_2] \cdots [g_\nu]$ of integral divisors, none of which is $[1]$. A process will be given which produces *either* an $[h]$ with the required properties *or* a proper refinement of the factorization of $[g]$, that is, a refinement in which at least one $[g_i]$ is given as a product of at least two integral divisors $\neq [1]$. Since the process never goes beyond the field extension K of r, of finite degree, generated by the coefficients of f and g, Corollary (5) of §1.17 shows that it must eventually produce an $[h]$.

By Theorem 1 and the Proposition of §1.19, the factorization of $[g]$ can be assumed to be of the form $[g] = [g_1]^{\sigma_1}[g_2]^{\sigma_2} \cdots$

$[g_\mu]^{\sigma_\mu}$, where the $[g_i]$ are relatively prime integral divisors $\neq [1]$ and the σ_i are positive integers. Let $[\widehat{g}] = [g_1][g_2]\cdots[g_\mu]$. For each $i = 1, 2, \ldots, \mu$, let h_i be a polynomial with coefficients in K such that $[h_i]$ is the product of $[f]$ and all the $[g_j]$ other than $[g_i]$; in other words, let $[h_i] = [f][\widehat{g}][g_i]^{-1}$. Then (Corollary (10), §1.12) $[f][\widehat{g}][g_i]^{-1}$ is the greatest common divisor of the coefficients of h_i. At least one of these coefficients must not be divisible by $[f][\widehat{g}]$ since otherwise $[f][\widehat{g}]|[f][\widehat{g}][g_i]^{-1}$, $[g_i]|[1]$, $[g_i] = [1]$, contrary to assumption. Thus, for each i, there is a $\beta^{(i)} \in K$ integral over r such that $[f][\widehat{g}][g_i]^{-1}$ divides $\beta^{(i)}$ but $[f][\widehat{g}]$ does not.

Again by Theorem 1, there are integral divisors $[\psi_1]$, $[\psi_2]$, $\ldots$, $[\psi_\tau]$ such that $[f]$, all the divisors $[g_i]$, and all the divisors $[\beta^{(i)}]$ are products of powers of the $[\psi]$'s. If one of the $[\psi]$'s is a proper divisor of $[g_i]$, then a refinement of the factorization of $[g]$ has been achieved. Otherwise, $[g_i]$ is *equal* to one of the $[\psi]$'s. In the expression of the integral divisor $[\beta^{(i)}]$ as $[f][\widehat{g}][g_i]^{-1}$ times a product of powers of the $[\psi]$'s, the product of powers of the $[\psi]$'s cannot include $[g_i]$ because $[f][\widehat{g}] \nmid [\beta^{(i)}]$. Therefore, $[f][g_i]$ does not divide $[\beta^{(i)}]$. Since the same is true for each i, if this procedure does not produce a refinement of the factorization of $[g]$, then $[f][g_i]$ does not divide $[\beta^{(i)}]$ for $i = 1, 2, \ldots, \mu$.

Assume $[f][g_i] \nmid [\beta^{(i)}]$ for $i = 1, 2, \ldots, \mu$, and set $\alpha = \beta^{(1)} + \beta^{(2)} + \cdots + \beta^{(\mu)}$. Let $[\phi_1]$, $[\phi_2]$, $\ldots$, $[\phi_\rho]$ be relatively prime integral divisors such that $[f]$, all $[g_i]$, $[\alpha]$, and all $[\beta^{(i)}]$ are products of powers of the $[\phi]$'s (Theorem 1). Again, if the $[\phi]$'s do not refine the factorization of $[g]$ then each $[g_i]$ must coincide with one of the $[\phi]$'s. Since $[f]$ divides each $[\beta^{(i)}]$, it divides $[\alpha]$, and $[\alpha] = [f]$ times a product of powers of the $[\phi]$'s. Since $[f][g_i]$ divides $\beta^{(j)}$ for $i \neq j$ but does not divide $\beta^{(i)}$, $[f][g_i]$ does not divide $\alpha = \sum \beta^{(j)}$ (Proposition (1) of §1.8). Thus, $[\alpha]$ is $[f]$ times a product of powers of $[\phi]$'s other than $[g_i]$. Since i was arbitrary and the $[\phi]$'s are relatively prime, $[\alpha] = [f][h]$, where $[h]$ is relatively prime to $[g]$, as was to be shown.

COROLLARIES. (1) *If $[f]$ is an integral divisor and if $\alpha \in K$ satisfies $[f]|[\alpha]$, then $[f] = [\alpha, \beta]$ for some $\beta \in K$.*

(2) *Let $[f]$ be an integral divisor in K and let $p \in r$ be an element of r which $[f]$ divides. Then there is a $\Psi \in K$ such that an element $\gamma \in K$ integral over r is divisible by $[f]^\mu$ if and only if $\gamma\Psi^\mu$ is divisible by p^μ.*

(When K is a number field, when $[f]$ is prime in K, and when p is prime in $r = \mathbf{Z}$, Corollary (2) gives Kummer's representation of "ideal prime factors" and their multiplicities in terms of the arithmetic of elements of K integral over $r = \mathbf{Z}$. See also Corollary 2, §2.11.)

DEDUCTIONS: (1) By the theorem, there is an integral divisor $[h]$ relatively prime to $[f]^{-1}[\alpha]$ such that $[f][h] = [\beta]$ for some $\beta \in K$. With $[g] = [f]^{-1}[\alpha]$, then, $[\alpha, \beta] = [fg, fh] = [f][g, h] = [f][1] = [f]$. (2) Let $[p] = [f][g]$. By the theorem, there is a $\Psi \in K$ such that $[\Psi] = [g][h]$ where $[f, h] = [1]$. Then $[f]^\mu|[\gamma]$ if and only if $[f]^\mu|[\gamma][h]^\mu$ ($[f, h] = [1]$ implies $[f^\mu, h] = [1]$ and $[f^\mu, h^\mu] = [1]$, which implies $[f^\mu, \gamma h^\mu] = [f^\mu, \gamma]$, by (18) and (17) of §1.15) which is true if and only if $[f]^\mu[g]^\mu|[\gamma][h]^\mu[g]^\mu$, that is, $[p]^\mu|[\gamma\Psi^\mu]$. By Proposition (7) of §1.9, this is true if and only if $p^\mu|\gamma\Psi^\mu$.

§1.21 The divisor class group.

A question of considerable importance in the theory of divisors is that of *determining whether a given divisor is principal*, where a principal divisor is by definition one of the form $[\alpha]$ with $\alpha \in K$. Otherwise stated, given a field $K \supset r$ and a divisor $[f]$ represented by a polynomial with coefficients in K, the problem is to determine whether $[f] = [\alpha]$ for some $\alpha \in K$. The answer depends on the field K—a divisor which is not principal in K may well be principal in some extension of K, although a divisor which is principal in K is principal in any extension of K.

This problem is closely related to the problem of *factorization in the ring R of elements of K integral over r*. Indeed, if $\gamma \in R$ can be factored $\gamma = \alpha\beta$, where α, $\beta \in R$ are not units of R, then $[\gamma] = [\alpha][\beta]$, which shows not only that $[\gamma]$ is reducible but also that it has factors which are *principal*. If $[\gamma]$ can be written as a product of factors prime in K, then the finite number of integral divisors in K which divide $[\gamma]$ can be listed, and γ factors in R if and only if there is a divisor in this list other than $[1]$ and $[\gamma]$ which is principal. (If $\alpha \in K$, and if $[\alpha]$ is a proper factor of $[\gamma]$, then $[\alpha]$ is integral and $\alpha \in R$; moreover, since $[\alpha]|\gamma$, $\beta = \gamma/\alpha \in R$. Finally, α and β are not units because their divisors are not $[1]$.)

A useful tool in the problem of determining which divisors are principal is the *divisor class group*, defined as follows: The multiplicative group of nonzero divisors represented by polynomials with coefficients in K was defined in §1.14. Let $\mathcal{D}$ denote this group. The nonzero principal divisors are a *subgroup* of $\mathcal{D}$, call it $\mathcal{D}_0$. The divisor class group is the quotient group $\mathcal{D}/\mathcal{D}_0$.

A *classification* of $\mathcal{D}$ is (1) a group G and (2) a homomorphism $\mathcal{D} \to G$ which is onto and has kernel $\mathcal{D}_0$. The problem of finding a classification *includes* the problem of determining whether a given divisor is principal (a given divisor is principal if and only if its image under the homomorphism is the identity of G) but putting the problem in this larger context can clarify it.

The divisor class group $\mathcal{D}/\mathcal{D}_0$ can also be defined as the semigroup of *integral* divisors modulo the subsemigroup of *principal integral* divisors, as can be seen as follows: Any element of $\mathcal{D}$ can be written $[f_1][f_2]^{-1}$, where $[f_1]$ and $[f_2]$ are integral. By Theorem 2, $[f_2][h] = [\alpha]$ for some integral divisor $[h]$ and for some $\alpha \in K$. Then $[f_1][f_2]^{-1}$ represents the same element of $\mathcal{D}/\mathcal{D}_0$ as $[f_1][f_2]^{-1}[\alpha] = [f_1][h]$ does. Thus, any class of $\mathcal{D}/\mathcal{D}_0$ can be represented by an integral divisor, and the natural map from the semigroup of integral divisors to $\mathcal{D}/\mathcal{D}_0$ is *onto* with kernel the subsemigroup of principal integral divisors, as was to be shown.

§1.22 Prime Factorization and Normal Extensions.

An interesting and useful topic in arithmetic is the study of the way in which divisors prime in K factor when K is extended. The uniqueness of factorization of divisors into primes (Corollary (2), §1.19) easily implies the following proposition about the factorization of primes in a *normal* extension.

PROPOSITION. *Let $K \supset r$ be an algebraic extension and let $L \supset K$ be a normal field extension of finite degree, that is, let the group G of automorphisms of L over r which leave all elements of K fixed be a finite group with the property that the only elements of L left fixed by all automorphisms in G are the elements of K. Let P be a divisor prime in L, and let P divide a divisor Q prime in K. Then Q is a product of divisors prime in L of the form*

$$Q = (P_1 P_2 \ldots P_\mu)^\nu$$

where the P_i are the distinct divisors of the form $S(P)$ for $S \in G$. Moreover, $\lambda \mu \nu = [L:K] =$ order of G, where λ is defined by *

$$\prod_{S \in G} S(P) = Q^\lambda.$$

PROOF: Here $S(P)$ means, of course, the divisor $[Sf]$, where f is any polynomial with coefficients in L which represents P. Because $[g]|[h]$ implies $[Sg]|[Sh]$, it is easily shown that $S(P)$ depends only on P and is prime in L. Since $P|Q$, and $S(Q) = Q$, $S(P)|Q$ for all $S \in G$. Therefore, $(\prod S(P))|Q^n$, where n is

---

*The divisor on the left side of this equation is the relative norm from L to K of P. See §3.17.

the order of G. If $P = [f]$, then $\prod S(P) = [\prod S(f)]$, which is a divisor represented by a polynomial with coefficientsin K (because the coefficients of $\prod S(f)$ are invariant under G). Since this divisor divides Q^n, and since Q is prime in K, Corollary (3) of §1.19 implies that $\prod S(P) = Q^\lambda$ for some integer $\lambda > 0$ ($\prod S(P)$ is divisible by P and is therefore not [1]). Since $\prod S(P)$ is a product of divisors prime in L, the same corollary implies $Q = P_1^{\nu_1} P_2^{\nu_2} \ldots P_\mu^{\nu_\mu}$ for some integers $\nu_1, \nu_2, \ldots, \nu_\mu$, all ≥ 0. Each P_i occurs in $\prod S(P)$ exactly n/μ times. Therefore, P_i divides the left side of $\prod S(P) = Q^\lambda$ exactly n/μ times, and divides the right side exactly $\lambda\nu_i$ times. Therefore, by the uniqueness of factorization into primes, $\nu_i = n/\mu\lambda$ for all i, as was to be shown.

Note that $P_1 P_2 \cdots P_\mu$ is invariant under G (because elements of G permute the P's), but is not a divisor in K when $\nu > 1$ (because its νth power is prime in K). Since $\nu > 1$ can occur†, divisors in L invariant under G need not be divisors in K. However, as is shown in §1.32, $\nu = 1$ for all but a *finite* number of divisors P prime in L, so most divisors in L invariant under G are in K.

To study factorization in extensions $L \supset K$ which are *not* normal, it is best to find a larger extension $L' \supset K$ which is normal and to apply the above proposition to the normal extensions $L' \supset L$ and $L' \supset K$.

§1.23 Rings of Values.

Given an algebraic extension K of a natural ring r, there is a way of associating to each nonzero integral divisor A in K a certain ring, which we will call the *ring of values of K at A*. Every nonzero divisor can be written as a quotient of integral divisors (§1.14). Since the greatest common divisor of numer-

---

†Let $r = \mathbf{Z}$, $K = k = \mathbf{Q}$, $L = \mathbf{Q}(\sqrt{2})$. Then $[\sqrt{2}]$ is invariant under G (G carries $\sqrt{2}$ to $\pm\sqrt{2}$) but is not a divisor in $\mathbf{Q}$ because its square is [2], which is prime in $\mathbf{Q}$.

ator and denominator can be cancelled, every nonzero divisor can be written as a quotient of *relatively prime* integral divisors. Since Corollary (1) of §1.19 implies that such a representation is unique (if $CD' = C'D$, where $[C, D] = [C', D'] = [1]$, then the factors of a common refinement of $CD' = C'D$ which divide C must divide C', hence $C|C'$, and, in the same way, $C'|C$, so $C = C'$, $D = D'$) a nonzero divisor has a well-defined *numerator* and *denominator*. We will say that an element $\alpha \in K$, $\alpha \neq 0$, is *finite at* A if the denominator of $[\alpha]$ is relatively prime to A. (Let the denominator of $[0]$ be considered to be $[1]$, so that 0 is finite at A for all A.) Then the elements of K finite at A form a *ring*; if α and β are both finite at A, then, by Theorem 2, there exist elements b_1 and b_2 of K integral over r and relatively prime to A such that $b_1\alpha$ and $b_2\beta$ are integral over r, say $\alpha = a_1/b_1$, $\beta = a_2/b_2$, which implies that the denominators of $[\alpha + \beta] = [(b_2a_1 + b_1a_2)/b_1b_2]$ and $[\alpha\beta] = [a_1a_2/b_1b_2]$ both divide $[b_1b_2]$ and are therefore relatively prime to A. We will say that an element $\alpha \in K$, $\alpha \neq 0$, is *zero at* A if A divides the numerator of $[\alpha]$ (which implies, of course, that α is finite at A). (Let the numerator of $[0]$ be considered to be $[0]$, so that 0 is zero at A for all A.) Then the elements of K which are zero at A are clearly an *ideal* in the ring of elements of K finite at A. Therefore, the quotient—elements finite at A modulo elements zero at A—is a well-defined *ring*. This ring is the *ring of values of K at A*.

§1.24.

PROPOSITIONS. *Let A, B, and C be nonzero integral divisors in K, and let V_A, V_B, and V_C be the rings of values of K at A, B, and C, respectively.*

(1) *If $A|B$, then there is a natural homomorphism $V_B \to V_A$ which is functorial in the sense that if $B|C$ then the homomorphism $V_C \to V_A$ which results from $A|C$ is the composition of the homomorphisms $V_C \to V_B \to V_A$ resulting from $B|C$ and $A|B$.*

(2) *The homomorphism of* (1) *is onto; it is one-to-one only if $A = B$.*

(3) (Chinese Remainder Theorem) *If A and B are relatively prime, then the maps $V_{AB} \to V_A$ and $V_{AB} \to V_B$ of* (1) *combine to give an isomorphism $V_{AB} \approx V_A \oplus V_B$, where $V_A \oplus V_B$ denotes ordered pairs, of an element of V_A and an element of V_B, added and multiplied componentwise.*

(4) *A is prime if and only if V_A is a field.* (Here $V_{[1]}$, which is a ring with one element, is not considered to be a field.)

(5) *As an additive group, the kernel of the homomorphism $V_B \to V_A$ of* (1) *is isomorphic to $V_{B/A}$. A specific isomorphism is given as follows. Choose $\beta \in K$ such that $[\beta] = AC$, where C is an integral divisor relatively prime to B. Let $\hat{\beta}$ be the element of V_B represented by β. The map $V_{B/A} \to V_B$ which sends an element of $V_{B/A}$ to its preimage under $V_B \to V_{B/A}$ times $\hat{\beta}$ is independent of the choice of preimage and gives such an isomorphism.*

PROOFS: (1) If α is finite at B and $A|B$, then α is finite at A. If α and α' are both finite at A and if $\alpha - \alpha'$ is zero at B, then $\alpha - \alpha'$ is zero at A. In other words, if α and α' represent the same elements of V_B, then they represent the same elements of V_A. Thus $V_B \to V_A$ can be defined to take the element of V_B represented by α to the element of V_A represented by α. The homomorphism of rings defined in this way clearly has the functorial property described in (1).

(2) By Theorem 1, if $A|B$, then $A = A_1|A_2,\ A_2|A_3,\ \ldots,$ $A_{\nu-1}|A_\nu = B$, where at each stage $A_i = C_i D_i^k$, $A_{i+1} = C_i D_i^{k+1}$, where C_i and D_i are relatively prime integral divisors. Thus, in order to prove (2), it will suffice to prove it in the special case $A = CD^k$, $B = CD^{k+1}$, $[C, D] = [1]$. By Theorem 2, there is an $\alpha \in K$ integral over r which is divisible by CD^k but not by CD^{k+1} (provided $D \neq [1]$). Therefore, α represents the zero class of V_A but not of V_B, which shows that $V_B \to V_A$ is not one-to-one. Let α represent a class of V_A, where $A = CD^k$. Then the denominator of $[\alpha]$ is relatively prime to CD^k. If $k > 0$, then α also represents a class of V_B,

where $B = CD^{k+1}$, and the image of this class of V_B is the given class of V_A. Thus, $V_B \to V_A$ is onto when $k > 0$. If $k = 0$, $V_B \approx V_A \oplus V_D$ by (3), which implies $V_B \to V_A$ is onto.

(3) If $[A, B] = [1]$, if $\alpha \in K$ is finite at AB, and if the images of α in both V_A and V_B are 0, then the numerator of $[\alpha]$ is divisible both by A and by B, which implies (by (16) of §1.15, $C = [1]C = [A, B]C = [AC, BC]$, so C is divisible by AB if it is divisible by both A and B) that the numerator of $[\alpha]$ is divisible by AB, that is, α represents the zero element of V_{AB}. Thus, $V_{AB} \to V_A \oplus V_B$ is one-to-one.

Let $\beta \in K$ be finite at A and let $\gamma \in K$ be finite at B. It is to be shown that there is an α, finite at AB, which represents the same element of V_A as β and the same element of V_B as γ. By Theorem 2, there exist $\phi, \psi \in K$, integral over r, such that $A|\phi$, $[B, \phi] = [1]$, $B|\psi$, $[A, \psi] = [1]$. By Theorem 1, $[\psi]$ and the denominator of $[\beta]$ can be written in the forms $D_1^{\sigma_1} D_2^{\sigma_2} \ldots D_\mu^{\sigma_\mu}$ and $D_1^{\tau_1} D_2^{\tau_2} \ldots D_\mu^{\tau_\mu}$, respectively, where the D_i are relatively prime integral divisors. The denominator of $[\beta \psi^n]$ then contains D_i to the power $\max(0, \tau_i - n\sigma_i)$. Thus, for n large, the denominator does not contain factors D_i which occur with nonzero exponent in $[\psi]$. Since $B|\psi$, it follows that $\beta \psi^n$ is finite at B, and therefore at AB. In the same way, $\gamma \phi^n$ is finite at AB for all sufficiently large n. Let

$$\alpha = \frac{\psi^n \beta + \phi^n \gamma}{\psi^n + \phi^n}.$$

Since ϕ^n is divisible by A and ψ^n is relatively prime to A, $\psi^n + \phi^n$ is relatively prime to A. In the same way, $\psi^n + \phi^n$ is relatively prime to B. Therefore, α is finite at AB for all sufficiently large n. Moreover, $\alpha - \beta = (\psi^n \beta + \phi^n \gamma - \psi^n \beta - \phi^n \beta)/(\psi^n + \phi^n) = \phi \cdot \phi^{n-1}(\gamma - \beta)/(\psi^n + \phi^n)$, which is ϕ times an element of K finite at A for large n. Therefore, $\alpha - \beta$ is zero at A for large n, that is, α and β represent the same elements of V_A. In the same way, α and γ represent the same elements of V_B for all sufficiently large n, so α has the required property when n is large.

(4) If α represents a nonzero element of V_A, then the denominator of $[\alpha]$ is relatively prime to A and its numerator is not divisible by A. If A is prime, the latter condition implies that the numerator of α is relatively prime to A; in this case, α^{-1} represents an element of V_A, which shows that the element of V_A represented by α has a multiplicative inverse. Conversely, if A is not prime, then either $A = [1]$ or A is reducible. If $A = [1]$, then V_A is not a field. If A is reducible, then $A = DE$, where neither D nor E is $[1]$. By Theorem 2, there exist β, $\gamma \in K$ integral over r such that $D|[\beta]$ with a quotient relatively prime to E, and $E|[\gamma]$ with a quotient relatively prime to D. Then β and γ represent nonzero elements of V_A whose product is the zero element of V_A, and V_A cannot be a field.

(5) Given an element of $V_{B/A}$, let γ be an element of K finite at B which represents it. The element of V_B represented by $\beta\gamma$ is independent of the choice of γ because if γ' is finite at B and represents the same element of $V_{B/A}$ then $\gamma - \gamma'$ is finite at B and zero at B/A, which implies that $\beta(\gamma - \gamma')$ is finite at B and zero at $A \cdot (B/A) = B$, that is, that $\beta\gamma$ and $\beta\gamma'$ represent the same element of V_B. This element is in the kernel of $V_B \to V_A$, because β is zero at A. If $\beta\gamma$ and $\beta\gamma'$ represent the same element of V_B, then the numerator of $[\beta(\gamma-\gamma')]$ is divisible by $B = A \cdot (B/A)$, which, by the choice of β, implies the numerator of $[\gamma-\gamma']$ is divisible by B/A, that is, γ and γ' represent the same element of $V_{B/A}$. Thus, the map $V_{B/A} \to V_B$ defined in this way is one-to-one into the kernel of $V_B \to V_A$. Finally, any element of the kernel of $V_B \to V_A$ can be represented by an element α of K finite at B and zero at A, say $[\alpha] = AD/E$, where D and E are relatively prime integral divisors and E is relatively prime to B. Then $\gamma = \alpha/\beta$ has divisor $[\gamma] = D/EC$ with denominator relatively prime to B, so γ represents an element of $V_{B/A}$ whose image is the given element of the kernel of $V_B \to V_A$, and the map is onto.

§1.25 Primitive Elements, Norms, and Traces.

The following method of describing extensions of finite degree is often useful.

PROPOSITION 1. *Let K be an extension of finite degree of a natural ring r. There is an irreducible, monic polynomial $F(x)$ in one indeterminate x with coefficients in r such that K is isomorphic to the simple algebraic extension $k(\alpha)$ of r obtained by adjoining a root α of F to the field of quotients k of r.*

PROOF: By the theorem of the primitive element [**E3**, p. 46], there is an $\alpha \in K$ which generates K over k. As was seen in §1.9, $h(\alpha) = 0$ for some nonzero, irreducible polynomial $h(x)$ with coefficients in r, say $b_0 \alpha^n + b_1 \alpha^{n-1} + \cdots + b_n = 0$, where $b_0 \neq 0$. Then $b_0 \alpha$ generates K over k and is a root of the monic polynomial $b_0^{n-1} h(x/b_0) = x^n + b_1 x^{n-1} + b_0 b_2 x^{n-2} + \cdots + b_0^{n-1} b_n = F(x)$, with coefficients in r. If $F(x)$ were reducible over k, then $h(x)$ would be reducible over k, which would imply (clear denominators) that $h(x)$ was reducible over r, contrary to assumption. Therefore, K is isomorphic to the field obtained by adjoining to k a root of the irreducible polynomial $F(x)$ [**E3**, p. 39].

PROPOSITION 2. *Let α be a generator of K over r which is integral over r, as in the previous proposition, and let $F(x)$ be a monic, irreducible polynomial with coefficients in r of which α is a root. Let L be a splitting field of F over k (the field of quotients of r), and let $F(x) = (x-\alpha)(x-\alpha')(x-\alpha'') \cdots (x - \alpha^{(n-1)})$ be the factorization of F over L. For any polynomial f with coefficients in K, $N_K f$ is equal to the product of the n polynomials $f = f^{(0)}, f^{(1)}, \ldots, f^{(n-1)}$ with coefficients in L obtained by expressing the coefficients of f as polynomials in α with coefficients in k, and substituting the n elements α, α', α'', $\ldots$, $\alpha^{(n-1)}$ of L in place of α in the resulting expression of f.*

PROOF: See [**E3**, p. 82].

The *trace* of a polynomial f with coefficients in an extension K of finite degree of a natural ring r is the coefficient of y^{n-1} in $N_K(y + f)$, where y is a new indeterminate. Otherwise stated, the trace of f with respect to the extension $K \supset r$, denoted $\operatorname{tr}_K(f)$, is the trace of the matrix which represents multiplication by f relative to a basis of K over k, the field of quotients of r. The first description of $\operatorname{tr}_K(f)$ shows that if $[f]$ is integral then $\operatorname{tr}_K(f)$ has coefficients in r (§1.17, Cor. (1)). The second description shows that if a, $b \in k$ and f, g are polynomials with coefficients in K then $\operatorname{tr}_K(af + bg) = a \operatorname{tr}_K(f) + b \operatorname{tr}_K(g)$.

PROPOSITION 3. *With the notation of Proposition 2, $\operatorname{tr}_K(f)$ is the* sum *of the n polynomials $f^{(0)}$, $f^{(1)}$, ..., $f^{(n-1)}$ with coefficients in L.*

PROOF: $N_K(y + f) = \prod_{i=0}^{n-1}(y + f^{(i)}) = y^n + y^{n-1}\sum f^{(i)} +$ terms of lower degree in y, by Proposition 2.

§1.26 Differents.

The notion of the *discriminant* of an algebraic extension K of a natural ring r plays an important role in Kronecker's theory of divisors, particularly in the study of the factorization of divisors of the form $[p]$, where p is a prime in r. Discriminants are also important, in the same way, in Dedekind's theory of ideals in algebraic number fields. However, Dedekind simplified and clarified the theory by introducing a more refined concept, a divisor whose *norm* is the discriminant*; he called the new divisor the *fundamental ideal*, but Hilbert renamed it the *different*, a name which has become standard. In Kroneckerian terms, the different (which Kronecker never used) can be described as follows.

---

*At least in the case $r = \mathbf{Z}$ considered by Dedekind, the discriminant is the norm of the different (see §2.9).

Given a subset $\alpha_1, \alpha_2, \ldots, \alpha_\nu$ of an extension K of finite degree of a natural ring r, its *different* with respect to K, denoted $\operatorname{dif}_K(\alpha_1, \alpha_2, \ldots, \alpha_\nu)$, is the divisor in K represented by the following polynomial with coefficients in K.

Let $u_1, u_2, \ldots, u_\nu$, and x be indeterminates, let $\bar\alpha = \alpha_1 u_1 + \alpha_2 u_2 + \cdots + \alpha_\nu u_\nu$, and let $F = N_K(x - \bar\alpha)$, a polynomial with coefficients in the field of quotients k of r, homogeneous of degree $n = [K:k]$ in $x, u_1, u_2, \ldots, u_\nu$. Let F be regarded as a polynomial $F(x)$ in x with coefficients in the ring $k[u_1, u_2, \ldots, u_\nu]$ of polynomials in the indeterminates $u_1, u_2, \ldots, u_\nu$ with coefficients in k. When h is a new indeterminate, $F(x + h) = F(x) + hF'(x) + h^2 G(x, h)$, where F' and G are uniquely determined polynomials with coefficients in $k[u_1, u_2, \ldots, u_\nu]$. Substitution of $\bar\alpha$ for x in $F'(x)$ gives a polynomial $F'(\bar\alpha)$ in $u_1, u_2, \ldots, u_\nu$ with coefficients in K. By definition, $\operatorname{dif}_K(\alpha_1, \alpha_2, \ldots, \alpha_\nu) = [F'(\bar\alpha)]$.

§1.27.

PROPOSITIONS. (1) *A set* $\alpha_1, \alpha_2, \ldots, \alpha_\nu \in K$ *generates* K *over* k *if and only if* $\operatorname{dif}_K(\alpha_1, \alpha_2, \ldots, \alpha_\nu) \neq [0]$.

(2) *For* $c \in k$ *and* $\alpha_1, \alpha_2, \ldots, \alpha_\nu \in K$, $\operatorname{dif}_K(c\alpha_1, c\alpha_2, \ldots, c\alpha_\nu) = [c]^{n-1} \operatorname{dif}_K(\alpha_1, \alpha_2, \ldots, \alpha_\nu)$, *where* $n = [K:k]$.

(3) *If* $\nu > 1$, $\operatorname{dif}_K(\alpha_1, \alpha_2, \ldots, \alpha_\nu)$ *divides* $\operatorname{dif}_K(\alpha_1, \alpha_2, \ldots, \alpha_{\nu-1})$.

(4) *If* $\alpha_1, \alpha_2, \ldots, \alpha_\nu$ *are integral over* r, *then* $\operatorname{dif}_K(\alpha_1, \alpha_2, \ldots, \alpha_\nu)$ *depends only on the ring* $r[\alpha_1, \alpha_2, \ldots, \alpha_\nu]$ *in* K *generated by the* α's *over* r; *in other words, if* $\beta_1, \beta_2, \ldots, \beta_\mu$ *can all be expressed as polynomials in the* α's *with coefficients in* r, *and all* α's *can be expressed as polynomials in the* β's *with coefficients in* r, *then* $\operatorname{dif}_K(\alpha_1, \alpha_2, \ldots, \alpha_\nu) = \operatorname{dif}_K(\beta_1, \beta_2, \ldots, \beta_\mu)$.

PROOFS: (1) By Proposition 2 of §1.25, $F(x) = \prod_{i=0}^{n-1}(x - \bar\alpha^{(i)})$, where $\bar\alpha^{(0)}, \bar\alpha^{(1)}, \ldots, \bar\alpha^{(n-1)}$ are the n conjugates of $\bar\alpha$ with coefficients in an extension L of K. Let $\bar\alpha = \bar\alpha^{(0)}$. Thus

$$F'(x) = \sum_{i=0}^{n-1}(x - \bar\alpha^{(0)}) \cdots (x - \bar\alpha^{(i-1)})(x - \bar\alpha^{(i+1)}) \cdots (x - \bar\alpha^{(n-1)})$$

from which it follows that $F'(\bar{\alpha}) = \prod_{i=1}^{n-1}(\bar{\alpha} - \bar{\alpha}^{(i)})$. If $F'(\bar{\alpha}) = 0$, then $\bar{\alpha} = \bar{\alpha}^{(i)}$ for some $i = 1, 2, \ldots, n-1$, that is, one of the conjugations $K \to L$ other than the identity carries all the coefficients $\alpha_1, \alpha_2, \ldots, \alpha_\nu$ of $\bar{\alpha}$ to themselves; therefore,* the α's cannot generate K over k. Conversely, if the α's do not generate K over k, they generate a proper subfield of K over k, call it K'. As follows easily from the definition of the norm, $N_K(x - \bar{\alpha}) = N_{K'}(x - \bar{\alpha})^{[K:K']}$, that is, $F(x) = F_0(x)^j$, where $F_0(x) = N_{K'}(x - \bar{\alpha})$ and $j = [K:K'] > 1$. Since $F(\bar{\alpha}) = 0$, $F_0(\bar{\alpha}) = 0$. Thus, $F'(\bar{\alpha}) = jF_0(\bar{\alpha})^{j-1}F_0'(\bar{\alpha}) = 0$, so $[F'(\bar{\alpha})] = [0]$, as was to be shown.

(Note that $[F'(\bar{\alpha})] = \prod_{i=1}^{n-1}[\bar{\alpha} - \bar{\alpha}^{(i)}]$ gives $\mathrm{dif}_K(\alpha_1, \alpha_2, \ldots, \alpha_\nu)$ as a product of divisors in L.)

(2) Replacing each α_i with $c\alpha_i$ in each of the $n-1$ factors of $F'(\bar{\alpha}) = \prod(\bar{\alpha} - \bar{\alpha}^{(i)})$ gives $c^{n-1}\prod(\bar{\alpha} - \bar{\alpha}^{(i)})$.

(3) If $F'(\bar{\alpha})$ is the polynomial which by definition represents $\mathrm{dif}_K(\alpha_1, \alpha_2, \ldots, \alpha_\nu)$, the polynomial which by definition represents $\mathrm{dif}_K(\alpha_1, \alpha_2, \ldots, \alpha_{\nu-1})$ is obtained by setting $u_\nu = 0$ in $F'(\bar{\alpha})$. Therefore, its coefficients are a subset of the coefficients of $F'(\bar{\alpha})$, so the divisor it represents is divisible by $\mathrm{dif}_K(\alpha_1, \alpha_2, \ldots, \alpha_\nu)$.

(4) Suppose $\beta = \psi(\alpha_1, \alpha_2, \ldots, \alpha_\nu)$, where ψ is a polynomial with coefficients in r. If S is an automorphism of L over k, then $\beta - S\beta = \psi(\alpha_1, \alpha_2, \ldots, \alpha_\nu) - \psi(S\alpha_1, S\alpha_2, \ldots, S\alpha_\nu)$ can be expressed as a linear combination of $\alpha_1 - S\alpha_1, \alpha_2 - S\alpha_2, \ldots, \alpha_\nu - S\alpha_\nu$ in which the coefficients are polynomials in the α's and the $S\alpha$'s and are therefore integral over r. Thus, for an indeterminate v, the coefficients of $(\bar{\alpha} + \beta v) - S(\bar{\alpha} + \beta v) = (\bar{\alpha} - S\bar{\alpha}) + (\beta - S\beta)v$ are all divisible by $[\bar{\alpha} - S\bar{\alpha}]$ (a divisor in L, namely, the g. c. d. of the $\alpha_i - S\alpha_i$). Therefore, $[(\bar{\alpha} + \beta v) - S(\bar{\alpha} + \beta v)] = [\bar{\alpha} - S\bar{\alpha}]$. Use this identity in $[F'(\bar{\alpha})] = \prod[\bar{\alpha} - S\bar{\alpha}]$ to find $\mathrm{dif}_K(\alpha_1, \alpha_2, \ldots, \alpha_\nu, \beta) = \mathrm{dif}_K(\alpha_1, \alpha_2, \ldots, \alpha_\nu)$. Thus,

---

*This argument is valid only when the extension is separable. (Note added in second printing.)

if the α's and β's are as in the statement of the proposition, $\operatorname{dif}_K(\alpha_1, \alpha_2, \ldots, \alpha_\nu) = \operatorname{dif}_K(\alpha_1, \alpha_2, \ldots, \alpha_\nu, \beta_1, \beta_2, \ldots, \beta_\mu) = \operatorname{dif}_K(\beta_1, \beta_2, \ldots, \beta_\mu)$.

Let $\alpha_1, \alpha_2, \ldots, \alpha_\nu \in K$ be integral over r, and let R_α denote $r[\alpha_1, \alpha_2, \ldots, \alpha_\nu]$, the subring of K generated by the α's and r. Proposition (4) shows that $\operatorname{dif}_K(\alpha_1, \alpha_2, \ldots, \alpha_\nu)$ depends only on R_α, not on the α's used to generate R_α. This divisor in K is the *different* of R_α relative to K, denoted $\operatorname{dif}_K(R_\alpha)$.

§1.28.

Let $\alpha_1, \alpha_2, \ldots, \alpha_\nu$ be integral over r in an extension K of r of finite degree, let $F'(\bar\alpha)$ be defined as in §1.26, and let R_α be as above.

PROPOSITION. *If h is a polynomial with coefficients in K, and if $[h]$ is integral, then $F'(\bar\alpha)h$ has coefficients in R_α.*

PROOF: It will suffice to prove the proposition in the case where h is constant—that is, h is an element of K—because a general h is a sum of multiples of such polynomials.

Given $h \in K$ integral over r, let $\operatorname{tr}_K(hq(x)) = \phi(x)$, where $q(x) = F(x)/(x-\bar\alpha)$. Since $F(x)$ has coefficients in r, the division algorithm shows that $q(x)$ has coefficients in the ring R_α. Therefore, the coefficients of $hq(x)$ are integral over r, which implies that the coefficients of $\phi(x)$ are in r. On the other hand, $\phi(x) = \sum \psi^{(i)}(x)$, where the $\psi^{(i)}(x)$ are the (not necessarily distinct) conjugates of $hq(x)$ in L (Proposition 3, §1.25). Therefore, $\phi(\bar\alpha) = \sum \psi^{(i)}(\bar\alpha)$. The summand $\psi^{(i)}(\bar\alpha)$ corresponding to the identity automorphism of L is $hq(\bar\alpha) = hF'(\bar\alpha)$. $(q(x) = (F(x) - F(\bar\alpha))/(x - \bar\alpha) = (F(\bar\alpha + x - \bar\alpha) - F(\bar\alpha))/(x - \bar\alpha) = (F(\bar\alpha) + (x-\bar\alpha)F'(\bar\alpha) + (x-\bar\alpha)^2 G(\bar\alpha, x - \bar\alpha) - F(\bar\alpha))/(x - \bar\alpha) = F'(\bar\alpha) + (x - \bar\alpha)G(\bar\alpha, x - \bar\alpha)$, so $q(\bar\alpha) = F'(\bar\alpha)$.) The other summands are all zero because they are the values at $x = \bar\alpha$ of h times $n - 1$ factors of the form $x - \bar\alpha^{(j)}$, one of which is $x - \bar{\alpha}$. Therefore, $hF'(\bar{\alpha}) = \phi(\bar{\alpha})$ has coefficients in R_α, as was to be shown.

COROLLARY 1. *If the different of R_α is not $[0]$, there is a divisor A in K such that all elements of K divisible by A are in R_α.*

DEDUCTION: By Proposition (1) of the preceding article, $\mathrm{dif}_K(R_\alpha) \neq [0]$ implies that any generator γ of K over k can be expressed as a polynomial in the α's with coefficients in k. Multiplication of such an expression by an element of r to clear denominators shows that a multiple of γ is in R_α. A multiple of γ is also a generator of K over k, so there is a generator γ of K over k in R_α. Then $\mathrm{dif}_K(R_\gamma) \neq [0]$ and R_γ, the subring of K generated by γ and r, is a subring of R_α, so it will suffice to prove the corollary in the case in which there is just one generator of the ring R_α. In this case, $F'(\bar{\alpha})$ is u_1^{n-1} times an element of K, call it $F'(\alpha)$. Let $A = [F'(\alpha)]$. If $A|\beta \in K$, then $\beta/F'(\alpha)$ is integral over r, and $\big(\beta/F'(\alpha)\big)F'(\bar{\alpha}) = \beta u_1^{n-1}$ is a polynomial with coefficients in R_α, as was to be shown.

A divisor A with the property of Corollary 1 is called a *conductor* of R_α in K. (*The** conductor is the greatest common divisor of the conductors.)

COROLLARY 2. *Given $\alpha_1, \alpha_2, \ldots, \alpha_\nu \in K$ integral over r generating K over k, there is an element Δ of r such that every element of K integral over r can be expressed in the form $\phi(\alpha)/\Delta$ where $\phi(\alpha) \in R_\alpha = r[\alpha_1, \alpha_2, \ldots, \alpha_\nu]$.*

DEDUCTION: Let A be as in Corollary 1 and let $\Delta \in r$ be divisible by A. If $\beta \in K$ is integral over r, then $\Delta\beta$ is divisible by A, so $\Delta\beta = \phi(\alpha)$, as was to be shown.

§1.29 Discriminants.

The *discriminant* with respect to K of a finite subset γ_1, $\gamma_2, \ldots, \gamma_\mu$ of K, denoted $\mathrm{disc}_K(\gamma_1, \gamma_2, \ldots, \gamma_\mu)$, is the fol-

---

*This definition requires that a proof of the g. c. d. of conductors is a conductor. (Note added in second printing.)

lowing divisor in K. Let S be the $\mu \times \mu$ symmetric matrix whose entry in the ith row of the jth column is $\mathrm{tr}_K(\gamma_i\gamma_j)$, an element of k. Let $n = [K\!:\!k]$ (finite by assumption), let U be an $n \times \mu$ matrix whose entries are indeterminates u_{ij}, and let U^t be the transpose of U. By definition*, $\mathrm{disc}_K(\gamma_1,\ \gamma_2, \ldots,\ \gamma_\mu) = [\det(USU^t)]$, the divisor represented by the polynomial $\det(USU^t)$ in the indeterminates u_{ij} with coefficients in k. The subscript K will sometimes be dropped from disc_K when there is no danger of confusion.

PROPOSITION. *If $\gamma_{\mu+1} = a_1\gamma_1 + a_2\gamma_2 + \cdots + a_\mu\gamma_\mu$, where the $a_i \in r$, then $\mathrm{disc}_K(\gamma_1,\gamma_2,\ldots,\gamma_{\mu+1}) = \mathrm{disc}_K(\gamma_1,\gamma_2,\ldots,\gamma_\mu)$.*

PROOF: Let S be the $\mu \times \mu$ symmetric matrix with entries $\big(\mathrm{tr}(\gamma_i\gamma_j)\big)$ for $1 \le i \le \mu$, $1 \le j \le \mu$, and let $\widehat{S}$ be the $(\mu+1) \times (\mu+1)$ symmetric matrix with entries $\big(\mathrm{tr}(\gamma_i\gamma_j)\big)$ for $1 \le i \le \mu+1$, $1 \le j \le \mu+1$. By the linearity of the trace, $\widehat{S} = MSM^t$, where M is the $(\mu+1) \times \mu$ matrix whose first μ rows are the $\mu \times \mu$ identity matrix and whose $(\mu+1)$st row is $a_1, a_2, \ldots, a_\mu$. Let V be an $n \times (\mu+1)$ matrix whose entries are indeterminates v_{ij}, let U be an $n \times \mu$ matrix whose entries are indeterminates u_{ij}, let $f = \det(USU^t)$, and let $g = \det(V\widehat{S}V^t)$. The coefficients of f are a *subset* of the coefficients of g; more precisely, f is obtained from g by the substitution $v_{ij} = u_{ij}$ for $j \le \mu$, $v_{i,\mu+1} = 0$. Therefore, $[g]\|[f]$. On the other hand, $g = \det(VMSM^tV^t)$, which is to say that g is obtained from f by the substitution $U = VM$, that is, $u_{ij} = v_{ij} + a_jv_{\mu+1,j}$. Since the a_j are in r, it follows that any common divisor of the coefficients of f divides all coefficients of g, that is, $[f]\|[g]$. Therefore, $[f] = [g]$, as was to be shown.

Corollary. *$\mathrm{Disc}_K(\gamma_1,\ \gamma_2,\ \ldots,\ \gamma_\mu)$ depends only on the r-module generated by the γ's. In other words, if $\gamma_1',\ \gamma_2',\ \ldots,\ \gamma_\nu' \in K$ satisfy $\gamma_i' = \sum a_{ij}\gamma_j$ and $\gamma_i = \sum b_{ij}\gamma_j'$, where the a_{ij}*

---

and b_{ij} are in r, then $\operatorname{disc}_K(\gamma_1, \gamma_2, \ldots, \gamma_\mu) = \operatorname{disc}_K(\gamma_1', \gamma_2', \ldots, \gamma_\nu')$.

DEDUCTION: By the proposition, both discriminants are equal to $\operatorname{disc}_K(\gamma_1, \gamma_2, \ldots, \gamma_\mu, \gamma_1', \gamma_2', \ldots, \gamma_\nu')$.

§1.30.

Any ring of the form $R_\alpha = r[\alpha_1, \alpha_2, \ldots, \alpha_\nu]$, where the $\alpha_i \in K$ are integral over r, is a finitely-generated r-module. Indeed, if τ is an integer such that, for each i, α_i^τ is a linear combination of lower powers of α_i with coefficients in r, then the τ^ν monomials γ of the form $\alpha_1^{e_1} \alpha_2^{e_2} \cdots \alpha_\nu^{e_\nu}$ with $0 \le e_i < \tau$ span R_α over r. By the above Corollary, the discriminant of these γ's depends only on R_α. It will be denoted $\operatorname{disc}_K(R_\alpha)$.

PROPOSITION. *As in §1.25, let $K = k(\alpha)$, where α is integral over r. Then the discriminant of $R_\alpha = r[\alpha]$ with respect to K is $N_K(\operatorname{dif}_K(\alpha))$.*

PROOF: As was seen in §1.27, $\operatorname{dif}_K(\alpha) = \prod[\alpha - \alpha^{(i)}]$, where $\alpha^{(1)}, \alpha^{(2)}, \ldots, \alpha^{(n-1)}$ are the roots, other than α, of $F(x) = N(x - \alpha)$ in a splitting field of F. Then, by Proposition 2 of §1.25, $N(\operatorname{dif}_K(\alpha)) = [N(\prod(\alpha - \alpha^{(i)}))] = [\prod(\alpha^{(j)} - \alpha^{(i)})]$, where the product is over all $n(n-1)$ pairs of roots $\alpha^{(j)}$, $\alpha^{(i)}$ of $F(x)$ in a splitting field ($\alpha = \alpha^{(0)}$, $n = [K : k]$) for which $\alpha^{(j)} \ne \alpha^{(i)}$. Therefore*, $N(\operatorname{dif}_K(\alpha)) = [\prod_{i<j}(\alpha^{(j)} - \alpha^{(i)})]^2$. Now $\prod_{i<j}(\alpha^{(j)} - \alpha^{(i)})$ is the determinant of the matrix whose ith column is 1, $\alpha^{(i+1)}$, $(\alpha^{(i+1)})^2$, $\ldots$, $(\alpha^{(i+1)})^{n-1}$ (a Vandermonde determinant). Let A denote this matrix. Then $N(\operatorname{dif}_K(\alpha)) = [\det(A)]^2 = [\det(AA^t)]$. With $\gamma_i = \alpha^{i-1}$, the entry in the ith row of the jth column of AA^t is $\operatorname{tr}(\gamma_i\gamma_j)$. Since the γ's span R_α, $\operatorname{disc}_K(R_\alpha) = [\det(UAA^tU^t)]$, where U is an $n \times n$ matrix

---

*This equation shows that $N(\operatorname{dif}_K(\alpha)) = \operatorname{disc}_K(R_\alpha)$ is the divisor represented by *the discriminant of the polynomial F* (see [E3, §13]), which is the reason for the name "discriminant".

of indeterminates. But $\det(UAA^tU^t) = \det(AA^t)\det(U)^2$ and $\det(U)$ is primitive (in fact its coefficients are all ± 1). Thus $[\det(U)] = [1]$ and $\operatorname{disc}_K(R_\alpha) = [\det(AA^t)] = N_K(\operatorname{dif}_K(\alpha))$, as was to be shown.

THEOREM. *Let M and M' be finitely generated r-modules in K, and let $M' \subset M$. Then $\operatorname{disc}_K(M') = A^2 \operatorname{disc}_K(M)$, where A is an integral divisor. In particular, if $\operatorname{disc}_K(M) = [0]$ then $\operatorname{disc}_K(M') = [0]$. If $\operatorname{disc}_K(M) \neq [0]$, then A can be found as follows. Let $\gamma_1, \gamma_2, \ldots, \gamma_\mu$ generate M and $\gamma'_1, \gamma'_2, \ldots, \gamma'_\nu$ generate M'. Choose a basis $\zeta_1, \zeta_2, \ldots, \zeta_n$ of K over k and let $T = (t_{ij})$, $T' = (t'_{ij})$ be the matrices with entries in k defined by $\gamma_i = \sum t_{ij}\zeta_j$, $\gamma'_i = \sum t'_{ij}\zeta_j$. Let U and V be $n \times \mu$ and $n \times \nu$ matrices, respectively, whose entries are indeterminates. Then $[\det(UT)] \neq [0]$ and $A = [\det(VT')][\det(UT)]^{-1}$.*

PROOF: Let $S_{\gamma'}$ be the $\nu \times \nu$ matrix $(\operatorname{tr}(\gamma'_i\gamma'_j))$ and let $S_{\gamma,\gamma'}$ be the analogous $(\mu + \nu) \times (\mu + \nu)$ matrix with $\gamma_1, \gamma_2, \ldots, \gamma_\mu, \gamma'_1, \gamma'_2, \ldots, \gamma'_\nu$ in place of $\gamma'_1, \gamma'_2, \ldots, \gamma'_\nu$. By definition, $\operatorname{disc}_K(M') = [f]$, where $f = \det(VS_{\gamma'}V^t)$. By the proposition of §1.29, $\operatorname{disc}_K(M) = [g]$, where $g = \det(WS_{\gamma,\gamma'}W^t)$, with W an $n \times (\mu + \nu)$ matrix of indeterminates. Since the coefficients of f are a subset of the coefficients of g, $[g]|[f]$. Thus, $\operatorname{disc}_K(M') = B\operatorname{disc}_K(M)$, where B is an integral divisor, and it is to be shown that B is a square. If $\operatorname{disc}_K(M) = [0]$, then $\operatorname{disc}_K(M') = [0]$, and B can be taken to be $[1]^2$. Assume $\operatorname{disc}_K(M) \neq [0]$, and let S_ζ be the $n \times n$ symmetric matrix $(\operatorname{tr}(\zeta_i\zeta_j))$. By the linearity of the trace, $S_{\gamma'} = T'S_\zeta(T')^t$. Thus, $\operatorname{disc}_K(M') = [\det(VT'S_\zeta(T')^tV^t] = [\det(S_\zeta)][\det(VT')]^2$ (because VT' is $n \times n$). In the same way, $\operatorname{disc}_K(M) = [\det(S_\zeta)][\det(UT)]^2$. Thus, $[\det(UT)] \neq [0]$ and $\operatorname{disc}_K(M') = [\det(S_\zeta)][\det(VT')]^2$ is $\operatorname{disc}_K(M)$ times the square of $A = [\det(UT)]^{-1}[\det(VT')]$. Therefore $A^2 = B$ is integral, which shows that A is integral and satisfies the required relation.

COROLLARY 1. *Let $a_1\gamma_1 + a_2\gamma_2 + \cdots + a_n\gamma_n + a_{n+1}\gamma_{n+1} = 0$,* *where the $a_i \in r$, the $\gamma_i \in K$ are integral over r, and γ_1, γ_2, ..., γ_n are a basis of K over k. Then*

$$[a_1, a_2, \ldots, a_{n+1}]^2 \operatorname{disc}_K(\gamma_1, \gamma_2, \ldots, \gamma_n)$$
$$= [a_{n+1}]^2 \operatorname{disc}_K(\gamma_1, \gamma_2, \ldots, \gamma_{n+1}).$$

In particular, if, as it is natural to assume, $[a_1, a_2, \ldots, a_{n+1}] = [1]$, then $[a_{n+1}]^2$ divides $\operatorname{disc}_K(\gamma_1, \gamma_2, \ldots, \gamma_n)$.

DEDUCTION: If $a_{n+1} = 0$, then $a_i = 0$ for all i and the equation is trivially true. Otherwise, in the theorem, let $\mu = n+1$, $\nu = n$, $\zeta_i = \gamma_i$ for $i = 1, 2, \ldots, n$, and $\gamma_i' = \gamma_i$ for $i = 1$, $2, \ldots, n$. It is to be shown that $A = [a_{n+1}]/[a_1, a_2, \ldots, a_{n+1}] = [\det(VT')]/[\det(UT)]$, that is, $[a_{n+1}][\det(UT)] = [a_1, a_2, \ldots, a_{n+1}][\det(VT')]$. Since T' is the $n \times n$ identity matrix, $[\det(VT')] = [\det(V)] = [1]$. The $(n+1) \times n$ matrix T has the $n \times n$ identity matrix in its first n rows and has $-a_1/a_{n+1}$, $-a_2/a_{n+1}$, ..., $-a_n/a_{n+1}$ in its last row. Let the indeterminates in the first n columns of the $n \times (n+1)$ matrix U be called x_{ij} and let the indeterminates in the last column be called w_1, w_2, ..., w_n. Then UT is the matrix which has $x_{ij} - (a_j/a_{n+1})w_i$ in the ith row of the jth column. Let M be the matrix obtained by multiplying the ith row of UT by w_1 and subtracting from it w_i times the first row, for $i = 2, 3, \ldots, n$. Then $\det(M) = w_1^{n-1} \det(UT)$, and M has $x_{1j} - (a_j/a_{n+1})w_1$ in the first row, $w_1 x_{ij} - w_i x_{1j}$ in rows after the first. It is clear from the definition of the determinant that the coefficients of $\det(M)$ are ± 1, $\pm a_1/a_{n+1}$, ..., $\pm a_n/a_{n+1}$. (For example, $-a_1/a_{n+1}$ is the coefficient of $w_1^n x_{22} x_{33} \ldots x_{nn}$ in $\det(M)$.) Thus, $[\det(UT)] = [a_1/a_{n+1}, a_2/a_{n+1}, \ldots, a_n/a_{n+1}, 1] = [a_1, a_2, \ldots, a_{n+1}]/[a_{n+1}]$, as was to be shown.

COROLLARY 2. *Let α_1, α_2, ..., $\alpha_\nu \in K$ be integral over r, and let $R_\alpha = r[\alpha_1, \alpha_2, \ldots, \alpha_\nu]$. Then $\operatorname{disc}_K(R_\alpha) \neq [0]$ if and only if the α's generate K over k.*

DEDUCTION: Let β be a generator of K over k which is integral over r (§1.25), let $R_\beta = r[\beta]$, and let $R_{\alpha,\beta} = r[\alpha_1, \alpha_2, \ldots, \alpha_\nu, \beta]$. By the theorem, $\mathrm{disc}_K(R_\beta) = A^2 \mathrm{disc}_K(R_{\alpha,\beta})$. By the proposition, $\mathrm{disc}_K(R_\beta) = N\big(\mathrm{dif}_K(R_\beta)\big)$, and by (1) of §1.27, $\mathrm{dif}_K(R_\beta) \neq [0]$. Therefore, $\mathrm{disc}_K(R_\beta) \neq [0]$, which implies $\mathrm{disc}_K(R_{\alpha,\beta}) \neq [0]$. Again by the theorem, $\mathrm{disc}_K(R_\alpha) = A_1^2 \mathrm{disc}_K(R_{\alpha,\beta})$, so $\mathrm{disc}_K(R_\alpha) = [0]$ if and only if $A_1 = [0]$, which, by the theorem, is true if and only if $\det(VT') = 0$, where V is an $n \times \nu$ matrix of indeterminates and T' is the $\nu \times n$ matrix whose rows give the coordinates of a set of elements which span R_α relative to a basis of K over k. If the α's fail to generate K over k, there is a basis $\zeta_1, \zeta_2, \ldots, \zeta_n$ of K over k whose first $s < n$ elements span the subspace of K generated over k by the α's; in this case, the last $n - s$ columns of T' are 0, $\det(VT') = 0$, and $\mathrm{disc}_K(R_\alpha) = [0]$. Conversely, if the α's do generate K over k, then the rows of T' span k^n, the rank of T' is n, a subset of n rows of T' has nonzero determinant, values in k (in fact in the set containing just 0 and 1) can be assigned to the indeterminates in V to make VT' have nonzero determinant, $\det(VT') \neq 0$, and $\mathrm{disc}_K(R_\alpha) \neq [0]$.

§1.31 Ramification.

DEFINITION. An element α integral over r *ramifies* in an algebraic extension K of r containing α if there is an element $\beta \in K$ not divisible by α whose square is divisible by α. (Note that ramification depends on the ambient field—an element which does not ramify in K may ramify in an extension of K, although if α ramifies in K, then it ramifies in any extension of K. Note also that the definition does not involve divisor theory.)

PROPOSITION. *An element $\alpha \in K$ integral over r ramifies in K if and only if $[\alpha]$ is divisible by the square of an integral divisor in K other than $[1]$.*

PROOF: If $A^2|[\alpha]$, where A is integral and $A \neq [1]$, then, by

Theorem 2, there is a $\beta \in K$ with $[\beta] = A^{-1}[\alpha]B$, where B is integral and relatively prime to A. Then $[\beta/\alpha] = A^{-1}B$ is not integral, so $\alpha \nmid \beta$, but $[\beta^2/\alpha] = A^{-2}[\alpha]B^2$ is integral, so $\alpha|\beta^2$.

Conversely, suppose $\alpha \nmid \beta$ but $\alpha|\beta^2$, where $\alpha \in K$ is integral over r. By Theorem 1, there exist relatively prime integral divisors $Q_1, Q_2, \ldots, Q_\mu$ such that both $[\alpha]$ and $[\beta]$ are products of powers of the Q_i, say $[\alpha] = \prod Q_i^{a_i}$ and $[\beta] = \prod Q_i^{b_i}$. Since $[\alpha]$ is integral, $a_i \geq 0$ for all i; since $[\alpha]|[\beta^2]$, $2b_i \geq a_i$ for all i; but since $[\alpha] \nmid [\beta]$, $b_j < a_j$ for at least one j (Proposition, §1.19). The inequalities $2a_j > 2b_j \geq a_j \geq 0$ imply first $2a_j > 0$, then $a_j > 0$, then $2b_j > 0$, then $b_j > 0$, then $b_j \geq 1$, then $2a_j > 2$, and finally $a_j > 1$. Therefore, $Q_j^2|[\alpha]$.

§1.32.

Let K be an extension of finite degree of a natural ring r. Let p be a prime element of r. The ring $V_{[p]}$ of values of K at $[p]$ contains the ring of values of k (the field of quotients of r) at $[p]$. This latter ring is a *field* ($[p]$ is prime in k), call it k_p, so $V_{[p]}$ is a *vector space* over the field k_p. We will say that K is *p-regular* if there is a basis of K over k whose elements are finite at $[p]$ and whose image in $V_{[p]}$ is a basis of $V_{[p]}$ over k_p. As will be shown in §2.7, when $r = \mathbf{Z}$, every K is p-regular for every p. I do not know of an example r, K, p in which K is not p-regular.

PROPOSITION 1. *Let K be p-regular and let* $[p] = C_1^{e_1} \cdot C_2^{e_2} \cdots C_\mu^{e_\mu}$, *where the C_i are relatively prime integral divisors. Let ι_p denote the natural map from elements of K finite at $[p]$ to $V_{[p]}$, and let f be a polynomial with coefficients in K finite at $[p]$. Then*

$$\iota_p(N_K f) = \prod_{i=1}^{\mu}(N_i f)^{e_i}$$

where $N_i f$ is the polynomial with coefficients in $k_p \subset V_{[p]}$ which is the determinant of the matrix representing multipli-

cation by f of elements of V_{C_i} relative to a basis of V_{C_i} over k_p.

PROOF: Let β_1, β_2, ..., β_n be a basis of K over k such that $\iota_p(\beta_1)$, $\iota_p(\beta_2)$, ..., $\iota_p(\beta_n)$ is a basis of $V_{[p]}$ over k_p (K is p-regular). Since the β's are a basis of K over k, the equations $f\beta_i = \sum f_{ij}\beta_j$ define polynomials f_{ij} with coefficients in k. Because k is the field of quotients of r, there is an $a \in r$ such that the polynomials af_{ij} all have coefficients in r. Moreover, one can assume $a \in r$ to be chosen with the additional property that either a is not divisible by p or at least one coefficient of one af_{ij} is not divisible by p (since otherwise a/p would have the same properties as a). If p divided a, then, because all coefficients of f, and all the β_i, are finite at $[p]$, $af\beta_i$ would be zero at $[p]$; then ι_p applied to $af\beta_i = \sum(af_{ij})\beta_j$ would give $0 = \sum \iota_p(af_{ij})\iota_p(\beta_j)$ and would imply, contrary to the choice of a, that $\iota_p(af_{ij}) = 0$ for all i and j (because the $\iota_p(\beta_j)$ are linearly independent over k_p). Therefore, $p \nmid a$ and the coefficients of $f_{ij} = af_{ij}/a$ are all finite at p. Thus, $\iota_p(f)\iota_p(\beta_i) = \sum \iota_p(f_{ij})\iota_p(\beta_j)$ and $\iota_p(Nf) = \iota_p(\det(f_{ij})) = \det(\iota_p(f_{ij}))$ is the determinant of the matrix which represents multiplication by $\iota_p(f)$ on $V_{[p]}$ relative to the basis $\iota_p(\beta_1)$, $\iota_p(\beta_2)$, ..., $\iota_p(\beta_n)$ of $V_{[p]}$ over k_p. Therefore, $\iota_p(Nf)$ can be computed using *any* basis of $V_{[p]}$ over k_p.

By the Chinese Remainder Theorem (§1.24, (3)), $V_{[p]}$ is a direct sum of subspaces $V_{C_i^{e_i}}$. Relative to a basis of $V_{[p]}$ which respects this direct sum decomposition—that is, a basis whose images under $V_{[p]} \to V_{C_i^{e_i}}$ are zero for all but one i—multiplication by f is represented by a block diagonal matrix, and the determinant of such a matrix is the product of the determinants of the matrices on the diagonal. Thus, $\iota_p(Nf) = \prod_{i=1}^{\mu} D_i$, where D_i is the determinant of the matrix which represents multiplication by f on $V_{C_i^{e_i}}$ relative to a basis of $V_{C_i^{e_i}}$ over k_p. It is to be shown that $D_i = (N_i f)^{e_i}$, where $N_i f$ is as in the proposition.

Let $\xi \in K$ be divisible by $C_i^{-1}[p]$ with a quotient relatively prime to C_i (Theorem 2). An element of $V_{[p]}$ has image 0 in V_{C_i} if and only if its product with $\iota_p(\xi)$ is zero. With respect to the basis $\iota_p(\beta_i)$ of $V_{[p]}$ over k_p, this gives a set of homogeneous linear conditions on k_p^n which describe the kernel of $V_{[p]} \to V_{C_i}$. Therefore, by linear algebra, one can find a subset of $V_{[p]}$ whose image in V_{C_i} is a basis of V_{C_i} over k_p. Let $\zeta_1, \zeta_2, \ldots, \zeta_\sigma \in K$ represent such a subset of $V_{[p]}$, so that the ζ's are finite at $[p]$ and their images in V_{C_i} are a basis over k_p. Let $\eta \in K$ be integral over r and have divisor $C_i Q$, where Q is integral and relatively prime to C_i. Then the images in $V_{C_i^{e_i}}$ of $\zeta_s \eta^t$, for $1 \le s \le \sigma$, $0 \le t < e_i$, is a basis of $V_{C_i^{e_i}}$ over k_p (by (5) of §1.24 and induction on e_i). The matrix which represents multiplication by f relative to this basis is a $\sigma e_i \times \sigma e_i$ matrix of $\sigma \times \sigma$ blocks, which is zero in blocks below the main diagonal and on the main diagonal is e_i copies of the $\sigma \times \sigma$ matrix whose determinant is $N_i f$. Therefore, the determinant is $(N_i f)^{e_i}$, as was to be shown.

PROPOSITION 2. *Let* $K = k(\alpha)$, *where* α *is integral over* r, *and let* $F(x) = N_K(x - \alpha)$. *If* $[p]^2$ *does not divide* $N[F'(\alpha)]$, *then* K *is p-regular. In fact,* $1, \alpha, \alpha^2, \ldots, \alpha^{n-1}$ *is a basis of* K *over* k *whose image in* $V_{[p]}$ *is a basis of* $V_{[p]}$ *over* k_p, *where* $n = [K : k]$.

PROOF: Let ι_p denote the natural map from elements of K finite at $[p]$ to $V_{[p]}$. The dimension of $V_{[p]}$ as a vector space over k_p is at most n, because, given any $n + 1$ elements of $V_{[p]}$, there exist elements $\gamma_1, \gamma_2, \ldots, \gamma_{n+1}$ of K finite at $[p]$ such that $\iota_p(\gamma_1), \iota_p(\gamma_2), \ldots, \iota_p(\gamma_{n+1})$ are the given elements, and, since $n = [K : k]$ and k is the field of quotients of r, there exist a_1, $a_2, \ldots, a_{n+1} \in r$ such that $a_1 \gamma_1 + a_2 \gamma_2 + \cdots + a_{n+1} \gamma_{n+1} = 0$; one can assume without loss of generality that $[a_1, a_2, \ldots, a_{n+1}] = [1]$, in which case $\iota_p(a_1)\iota_p(\gamma_1) + \iota_p(a_2)\iota_p(\gamma_2) + \cdots + \iota_p(a_{n+1})\iota_p(\gamma_{n+1}) = 0$ and the $\iota_p(a_i)$ are not all zero, which shows that the given classes are linearly dependent over k_p.

The proposition will therefore be proved if it is shown that $\iota_p(\alpha^i)$ for $i = 0, 1, \ldots, n-1$ are linearly independent over k_p.

If the $\iota_p(\alpha^i)$ were linearly dependent over k_p, there would be elements c_i and d_i of r for $i = 0, 1, \ldots, n-1$ such that $\sum_{i=0}^{n-1} \iota_p(c_i/d_i)\iota_p(\alpha^i) = 0$, such that no d_i was divisible by p, and such that at least one c_i was not divisible by p, say $p \nmid c_r$. Multiplication by the product of the $\iota_p(d_i)$ would then give $\iota_p(t_0 + t_1\alpha + \cdots + t_{n-1}\alpha^{n-1}) = 0$, where the t_i were in r and $p \nmid t_r$. If $\beta = t_0 + t_1\alpha + \cdots + t_{n-1}\alpha^{n-1}$, then β would be integral over r and $\iota_p(\beta) = 0$, which would imply $\beta = p\gamma$, where γ was integral over r. By Corollary 1 of §1.30, $p\gamma = t_0 + t_1\alpha + \cdots + t_{n-1}\alpha^{n-1}$ would imply $[p]^2 \mid \operatorname{disc}_K(1, \alpha, \ldots, \alpha^{n-1}) = N[F'(\alpha)]$ (§1.30, Proposition), contrary to assumption. Therefore, the $\iota_p(\alpha^i)$ are linearly independent.

PROPOSITION 3. *Let $K \supset r$ be p-regular, and let $\alpha_1, \alpha_2, \ldots,$ α_ν be integral over r and generate K over k. If $[p, \operatorname{dif}_K(R_\alpha)] = [1]$ then p does not ramify in K.*

PROOF: If p ramifies in K, then $[p] = A^2 B$, where A and B are integral divisors and $A \neq [1]$. By Theorem 1, $[p] = C_1^{e_1} C_2^{e_2} \cdots C_\mu^{e_\mu}$, where the C_i are relatively prime integral divisors and A is a product of powers of the C_i. In particular, at least one C_i divides A, say $C_1 \mid A$. Then $e_1 \geq 2$.

Let $F(x, u_1, u_2, \ldots, u_\nu) = N_K(x - \bar\alpha)$, where $\bar\alpha = \alpha_1 u_1 + \alpha_2 u_2 + \cdots + \alpha_\nu u_\nu$. By Proposition 1, $\iota_p(F) = \prod_{i=1}^{\mu} g_i^{e_i}$, where g_i is the determinant of the matrix which represents multiplication by $x - \bar\alpha$ relative to a basis of V_{C_i} over k_p. The argument of §1.17 generalizes to prove that g_i is divisible by $x - \bar\alpha$ when both are regarded as polynomials with coefficients in V_{C_i}. Specifically, let $\zeta_1, \zeta_2, \ldots, \zeta_\sigma$ be a basis of V_{C_i} over k_p, and let M be the $\sigma \times \sigma$ matrix of polynomials with coefficients in k_p which represents multiplication of elements of V_{C_i} by $x - \bar\alpha$ relative to this basis. Then $g_i = \det(M)$. On the other hand, substitution of $\bar\alpha$ for x in M gives a matrix $\overline{M}$ of polynomials with coefficients in K such that $\overline{M}\zeta = 0$ when ζ

is the column matrix $\zeta_1, \zeta_2, \ldots, \zeta_\sigma$ and when the coefficients of entries of $\overline{M}$ are regarded as representing elements of V_{C_i}. Therefore, $\det(\overline{M})$ is zero as a polynomial with coefficients in V_{C_i}, or, what is the same, g_i is divisible by $x - \bar{\alpha}$ when both are regarded as polynomials with coefficients in V_{C_i}. Therefore, $(x - \bar{\alpha})^2$ divides F when both are regarded as polynomials with coefficients in $V_{C_1} \supset k_p$. In other words, the remainder when $N(x - \bar{\alpha})/(x - \bar{\alpha})$ is divided by $x - \bar{\alpha}$ has all coefficients divisible by C_1, which is precisely what it means to say that $C_1 | \operatorname{dif}_K(\alpha_1, \alpha_2, \ldots, \alpha_\nu)$. Since $C_1 | [p]$ and $C_1 \neq [1]$, it follows that $[p, \operatorname{dif}_K(R_\alpha)] \neq [1]$, as was to be shown.

COROLLARY 1. *In an extension K of r of finite degree, at most a finite number of primes ramify.*

DEDUCTION: Let $K = k(\alpha)$, where α is integral over r. By Proposition 2, K is p-regular for all but a finite number of primes p in r. By Proposition 3, of the primes p for which K is p-regular, p ramifies for at most a finite number.

COROLLARY 2. *In the proposition of §1.22, for fixed L and K, ν is greater than 1 for at most a finite number of P. In other words, if P is a divisor prime in L which divides a divisor prime in K, then, with at most a finite number of exceptions, the product of the distinct conjugates of P over K is a divisor in K. (As §1.22 shows, if the product of the distinct conjugates of P over K is a divisor in K, it is prime in K, because ν is the smallest power of this divisor which is in K.)*

DEDUCTION: If $\nu > 1$, then $P^2 | Q$ and $Q | [p]$ shows that p is in the finite set of primes which ramify, so P is in the finite set of divisors which divide such $[p]$.

Part 2: Applications to Algebraic Number Theory

§2.1 Factorization into Primes.

Let $\mathbf{Z}$ denote the ring of integers. An *algebraic number field* is an extension of $\mathbf{Z}$ of finite degree. Since $\mathbf{Z}$ is a natural ring, divisor theory applies to algebraic number fields.

THEOREM. *Let K be an algebraic number field and let $p \in \mathbf{Z}$ be a prime integer. The factorization of $[p]$ into divisors prime in K can be accomplished in a finite number of steps.*

LEMMA. *There is a finite set $\gamma_1, \gamma_2, \ldots, \gamma_\mu$ of elements of K integral over $\mathbf{Z}$ such that every element of K integral over $\mathbf{Z}$ is of the form $c_1\gamma_1 + c_2\gamma_2 + \cdots + c_\mu\gamma_\mu$ for $c_1, c_2, \ldots, c_\mu \in \mathbf{Z}$.*

PROOF: Let $\mathbf{Q}$ denote the field of quotients of $\mathbf{Z}$, the field of rational numbers. Because $[K\colon\mathbf{Q}]$ is finite, $K = \mathbf{Q}(\alpha)$ for some α integral over $\mathbf{Z}$ (§1.25). By the second corollary of §1.28, there is a $\Delta \in \mathbf{Z}$ such that every element of K integral over $\mathbf{Z}$ is of the form $(a_0 + a_1\alpha + a_2\alpha^2 + \cdots + a_{n-1}\alpha^{n-1})/\Delta$. Since each a_i can be written in the form $a_i = q_i\Delta + r_i$, where $|r_i| \leq |\Delta|/2$, every element of K integral over $\mathbf{Z}$ is of the form $q_0 + q_1\alpha + q_2\alpha^2 + \cdots + q_{n-1}\alpha^{n-1} + \delta$, where δ is one of the finite set of elements of K of the form $(\sum_{i=0}^{n-1} s_i\alpha^i)/\Delta$, with integers s_i in the range $2|s_i| \leq |\Delta|$, and is integral over $\mathbf{Z}$. If $\gamma_i = \alpha^{i-1}$ for $i = 1, 2, \ldots, n$, and $\gamma_{n+1}, \gamma_{n+2}, \ldots, \gamma_\mu$ is the list of elements of the form $(\sum s_i\alpha^i)/\Delta$ with the s_i in this range which are integral over $\mathbf{Z}$, then every element of K integral over $\mathbf{Z}$ is of the form $\sum_{i=1}^{n} q_i\gamma_i + \gamma_j$ for some j in the range $n < j \leq \mu$, and the lemma follows.

PROOF OF THE THEOREM: By Corollary (1) of §1.20, every integral divisor of $[p]$ in K can be written in the form $[p, \beta]$ for some $\beta \in K$. Since β is integral over $\mathbf{Z}$ (Cor. (13), §1.12), $\beta = \sum_{i=1}^{\mu} c_i\gamma_i$, where the γ_i are as in the lemma and the $c_i \in \mathbf{Z}$. Clearly, $[p, \beta] = [p, \beta \pm p\gamma_i]$ for any γ_i, from which it follows that $[p, \beta] = [p, \sum_{i=1}^{\mu} b_i\gamma_i]$ where $\sum_{i=1}^{\mu} b_i\gamma_i$ is one of the p^μ elements* of K of this form in which the b_i are integers satisfying $0 \le b_i < p$. Therefore, the p^μ divisors $[p, \sum_{i=1}^\mu b_i \gamma_i]$ include *all* integral divisors of $[p]$. Given any factorization of $[p]$ (counting $[p]$ itself as a factorization in which there is one factor) one can either refine it further (if one of the p^μ divisors $[p, \sum_{i=1}^\mu b_i \gamma_i]$ is a proper divisor of one of its factors) or prove there is no refinement (if not). Since the number of integral factors of $[p]$ other than $[1]$ is bounded (§1.17, Corollary (5)) the theorem follows.

COROLLARY. *Any divisor in K can be written as a product of powers (possibly negative) of distinct divisors prime in K.*

DEDUCTION: By the theorem, this is true of divisors of the form $[p]$. Therefore, it is true of any product of divisors of this form, that is, any divisor of the form $[m]$, where $m \in \mathbf{Z}$. Corollary (3) of §1.19 then implies it is true of any integral divisor which divides a divisor of the form $[m]$, which is to say, by Lemma (8) of §1.10, any integral divisor. Since every divisor is a quotient of integral divisors, the general case follows.

§2.2 A Factorization Method.

The factorization of $[p]$ into divisors prime in a given algebraic number field K can *usually* be found as follows. $K = \mathbf{Q}(\alpha)$ where α is integral over $\mathbf{Z}$ (§1.25). Let $F(x)$ be $N_K(x - \alpha)$, a monic, irreducible polynomial of degree $n = [K:\mathbf{Q}]$ with coefficients in $\mathbf{Z}$. Let the factorization of $F(x)$ into powers of distinct, irreducible factors mod p be $F(x) \equiv f_1(x)^{e_1} f_2(x)^{e_2} \ldots f_m(x)^{e_m} \bmod p$. Then

(1) $[p] = [p, f_1(\alpha)]^{e_1} [p, f_2(\alpha)]^{e_2} \ldots [p, f_m(\alpha)]^{e_m}$

---

*It is at this point that the assumption $r = \mathbf{Z}$ enters. Kronecker claims, with an indication of a proof [**Krl**, §6], that the above lemma is true for r a ring of polynomials with integer coefficients, which implies that every divisor of $[p]$ is of the form $[p, \sum b_i \gamma_i]$ where the $b_i \in r$ are reduced mod p. When $r = \mathbf{Z}$, the set of such divisors is *finite*.

is usually the factorization of $[p]$, that is, usually the divisors $[p, f_i(\alpha)]$ are *distinct* and *prime in* K, and (1) holds.

(The precise definition of the divisors $[p, f_i(\alpha)]$ requires a little explanation. Strictly speaking, the coefficients of $f_i(x)$ are in $\mathbf{F}_p$, the field of integers mod p—or, better, the ring of values of $\mathbf{Q}$ at the divisor $[p]$ prime in $\mathbf{Q}$—and $f_i(\alpha)$ is not meaningful. However, it is meaningful to say that $\widehat{f}_i(x) \equiv f_i(x) \bmod p$, where $\widehat{f}_i(x)$ is a polynomial with coefficients in $\mathbf{Z}$, and then $\widehat{f}_i(\alpha)$ and $[p, \widehat{f}_i(\alpha)]$ are well defined. If $g(x)$ and $h(x)$ are polynomials with coefficients in $\mathbf{Z}$, and if $g(x) \equiv h(x) \bmod p$, then $g(\alpha) - h(\alpha)$ is divisible by p—because α is integral over $\mathbf{Z}$—and $[p, g(\alpha)] = [p, g(\alpha), g(\alpha) - h(\alpha)] = [p, h(\alpha)]$. Therefore, $[p, \widehat{f}_i(\alpha)]$ is independent of the choice of $\widehat{f}_i(x)$. This is the divisor denoted $[p, f_i(\alpha)]$ in (1).)

§2.3.

The method of factoring $[p]$ in K sketched in the preceding article depends on the choice of a generator α of K over $\mathbf{Q}$, one which is integral over $\mathbf{Z}$. As examples show (see §2.5), the method may fail to factor $[p]$ for one choice of α and succeed for another. However, there are algebraic number fields K and primes p for which the method fails, no matter which generator α is used (§2.5). Kronecker strengthened the method by replacing α with a linear combination $\bar{\alpha} = \alpha_1 u_1 + \alpha_2 u_2 + \cdots + \alpha_\nu u_\nu$ of elements α_i of K integral over $\mathbf{Z}$ with indeterminate coefficients u_i. Then $F(x) = N(x - \alpha)$ is replaced by $F(x, u_1, u_2, \ldots, u_\nu) = N(x - \bar{\alpha}) = N(x - \alpha_1 u_1 - \alpha_2 u_2 - \cdots - \alpha_\nu u_\nu)$, a homogeneous polynomial of degree $[K : \mathbf{Q}]$ in $x, u_1, u_2, \ldots, u_\nu$ with coefficients in $\mathbf{Z}$. This F can be factored mod p, that is, written in the form $F \equiv f_1^{e_1} f_2^{e_2} \cdots f_m^{e_m} \bmod p$, where the f_i are distinct polynomials in $x, u_1, u_2, \ldots, u_m$ (necessarily homogeneous) with coefficients in $\mathbf{F}_p$, irreducible over $\mathbf{F}_p$. The theorem of the next article gives conditions on $\alpha_1, \alpha_2, \ldots, \alpha_\nu$ which imply that the factorization of $[p]$ is

(1) $$[p] = [p, f_1(\bar{\alpha})]^{e_1} [p, f_2(\bar{\alpha})]^{e_2} \cdots [p, f_m(\bar{\alpha})]^{e_m}$$

where $[p, f_i(\bar{\alpha})]$ denotes the following divisor in K: Let $\widehat{f}_i$ be a polynomial with coefficients in $\mathbf{Z}$ such that $\widehat{f}_i \equiv f_i \bmod p$. Substitution of $\bar{\alpha} = \alpha_1 u_1 + \alpha_2 u_2 + \cdots + \alpha_\nu u_\nu$ for x in $\widehat{f}_i$ gives a polynomial, call it $\widehat{f}_i(\bar{\alpha})$, in $u_1, u_2, \ldots, u_\nu$ whose coefficients are in K. The divisor $[p, f_i(\bar{\alpha})]$ is by definition $[p, \widehat{f}_i(\bar{\alpha})]$, that is, the greatest common divisor of p and the coefficients of $\widehat{f}_i(\bar{\alpha})$. This definition is valid because different choices of $\widehat{f}_i$ give the same divisor $[p, \widehat{f}_i(\bar{\alpha})]$. (The α_i are integral over $\mathbf{Z}$, so $g(x) \equiv h(x) \bmod p$ implies that all coefficients of $g(\bar{\alpha}) - h(\bar{\alpha})$ are divisible by p.)

§2.4.

DEFINITION. A ring R contained in an algebraic number field K *absorbs division by an integer* $j \in \mathbf{Z}$ if $\phi \in R$ and $j | \phi$ imply $\phi/j \in R$.

THEOREM. *Let* $K = \mathbf{Q}(\alpha_1, \alpha_2, \ldots, \alpha_\nu)$ *be an algebraic number field generated over* $\mathbf{Q}$ *by a finite number of elements* α_i *integral over* $\mathbf{Z}$, *and let* $R_\alpha = \mathbf{Z}[\alpha_1, \alpha_2, \ldots, \alpha_\nu]$ *be the ring generated in* K *by* $1, \alpha_1, \alpha_2, \ldots, \alpha_\nu$. *Let* $\bar{\alpha}$ *be* $\alpha_1 u_1 + \alpha_2 u_2 + \cdots + \alpha_\nu u_\nu$, *a linear polynomial in the indeterminates* $u_1, u_2, \ldots, u_\nu$ *with coefficients in* K, *and let* F *be* $N(x - \bar{\alpha})$, *a homogeneous polynomial of degree* $[K:\mathbf{Q}]$ *in* $x, u_1, u_2, \ldots u_\nu$, *monic in* x, *with coefficients in* $\mathbf{Z}$. *Finally, let* $F \equiv f_1^{e_1} f_2^{e_2} \cdots f_m^{e_m} \bmod p$ *be the factorization of* F *over* $\mathbf{F}_p$ *as a product of powers of distinct irreducible monic factors. If* R_α *absorbs division by* p *then*

(1) $$[p] = [p, f_1(\bar{\alpha})]^{e_1} [p, f_2(\bar{\alpha})]^{e_2} \cdots [p, f_m(\bar{\alpha})]^{e_m}$$

gives $[p]$ *as a product of powers of distinct divisors prime in* K, *where* $[p, f_i(\bar{\alpha})]$ *is the divisor described in the preceding article.*

COROLLARIES. (1) *The method of §2.2 gives the factorization of* $[p]$ *whenever* $[p]^2$ *does not divide* $N[F'(\alpha)]$.

(2) *The method of §2.2 gives the factorization of* $[p]$ *whenever the exponents* e_i *are all 1.*

DEDUCTIONS: (1) If $R_\alpha = \mathbf{Z}[\alpha]$ does not absorb division by p, then some $\beta \in K$ of the form $\beta = (a_0 + a_1\alpha + \cdots + a_{n-1}\alpha^{n-1})/p$, in which the a_i are integers not all divisible by p, is integral over $\mathbf{Z}$. By Corollary 1 of §1.30, $[p]^2$ then divides disc(1, α, ..., α^{n-1}) = disc(R_α). By the proposition of §1.30, disc(R_α) = $N[F'(\alpha)]$. (2) If the method of §2.2 fails, then $[p]^2 | N[F'(\alpha)]$. By Proposition 2 of §1.25, $N[F'(\alpha)]$ is the product of $n = [K:\mathbf{Q}]$ conjugates of $[F'(\alpha)]$ in a normal extension L of $\mathbf{Q}$ containing K. A prime divisor of $[p]$ in L must divide at least one of these conjugates of $[F'(\alpha)]$, so some conjugate of it must divide $[F'(\alpha)]$ itself, say P is a divisor of $[p]$ prime in L which divides $F'(\alpha)$. In the field $V_P \supset \mathbf{F}_p$ of values of L at P, the polynomials $F(x)$ and $F'(x)$ have the common root $\iota(\alpha)$, the element of V_P represented by α. Two polynomials in one indeterminate x with coefficients in $\mathbf{F}_p$ have a common root in an extension of $\mathbf{F}_p$ if and only if they have a common factor with coefficients in $\mathbf{F}_p$. Since $F \equiv \prod_i f_i^{e_i}$, it follows that some f_i divides F', say $f_1 | F'$ over $\mathbf{F}_p$. Since $F' \equiv e_1 f_1' f_1^{-1} F + $ (terms divisible by f_1), it follows that f_1 divides $e_1 f_1' f_1^{e_1-1} f_2^{e_2} f_3^{e_3} \cdots$. Since $f_1 \nmid f_i$ for $i > 1$ (unique factorization over $\mathbf{F}_p$) and $f_1 \nmid f_1'$ ($f_1 | f_1'$ would imply $f_1' = 0$ because f_1' has lower degree than f_1, and $f_1' = 0$ would imply f_1 was a pth power by Fermat's theorem, contrary to the assumption that f_1 is irreducible), it follows that $f_1 | e_1 f_1^{e_1-1}$. Therefore, $e_1 > 1$.

Corollary (1) shows that the method of §2.2 fails for at most a finite number of primes p, and justifies the assertion of §2.2 that the factorization method there "usually" works.

Note that, when $\alpha_1, \alpha_2, \ldots, \alpha_\nu$ have the property of $\gamma_1, \gamma_2, \ldots, \gamma_\mu$ in the lemma of §2.1, the ring R_α is the set of *all* elements of K integral over $\mathbf{Z}$, and this ring obviously absorbs division by p for all p. In this case, therefore, factorization of $N(x - \bar{\alpha})$ mod p gives the factorization of $[p]$ in K for *all* primes $p \in \mathbf{Z}$.

§2.5 Examples.

A quadratic extension K of $\mathbf{Q}$ is one which can be obtained by adjoining a root of an irreducible polynomial $a_0 x^2 + a_1 x + a_2$ with $a_i \in \mathbf{Z}$. If α is a root of this equation in K, then $(2a_0\alpha)^2 + 2a_1(2a_0\alpha) + 4a_0 a_2 = 0$ and $\beta = 2a_0\alpha + a_1$ satisfies $(\beta - a_1)^2 + 2a_1(\beta - a_1) + 4a_0 a_2 = 0$, which has the form $\beta^2 = d$ for $d \in \mathbf{Z}$. Since $K = \mathbf{Q}(\sqrt{k^2 c}) = \mathbf{Q}(\sqrt{c})$, one can assume without loss of generality that d is divisible by no square greater than 1. Since $K \neq \mathbf{Q}$, d is neither 0 nor 1. $F(x) = x^2 - d$, $F'(x) = 2x$, $F'(\sqrt{d}) = 2\sqrt{d}$, so $N[F'(\sqrt{d})] = [2]^2[d]$. Since d is squarefree, Corollary (1) of the preceding article shows that the method of §2.2 gives the factorization of $[p]$ in $\mathbf{Q}(\sqrt{d})$ for all primes p except, possibly, $p = 2$. The factorization of $[p]$ in $\mathbf{Q}(\sqrt{d})$ is found by factoring $N(x - \sqrt{d}) = x^2 - d$ mod p, that is, by finding the roots, if any, of $x^2 \equiv d$ mod p. If this congruence has no root, then $[p]$ is prime in $\mathbf{Q}(\sqrt{d})$, if it has two distinct roots $\pm c$ mod p, then $[p]$ is $[p, \sqrt{d} - c][p, \sqrt{d} + c]$, a product of distinct divisors prime in $\mathbf{Q}(\sqrt{d})$, and if it has just one root c mod p (which must be 0 mod p unless $p = 2$) then $[p] = [p, \sqrt{d} - c]^2$.

Since $d^2 \equiv d$ mod 2, this method gives the factorization $[2] = [2, \sqrt{d} - d]^2$ of $[2]$ *provided* $R_{\sqrt{d}}$ absorbs division by 2, that is, provided $2 | (a + b\sqrt{d})$, for $a, b \in \mathbf{Z}$, implies $2 | a$ and $2 | b$. Now $2 | (a + b\sqrt{d})$ implies $4 | N(a + b\sqrt{d}) = (a^2 - db^2)$. If $2 | a$, then $4 | (a^2 - db^2)$ implies $4 | db^2$, and hence, since $4 \nmid d$, implies $2 | b^2$ and $2 | b$. If $2 \nmid a$, then $a^2 \equiv 1$ mod 4, and $4 | (a^2 - db^2)$ implies $db^2 \equiv 1$ mod 4, which implies $2 \nmid b$ and $d \equiv 1$ mod 4. Thus, *if* $R_{\sqrt{d}}$ *does not absorb division by 2, then* $d \equiv 1$ mod 4.

It remains to factor $[2]$ in $\mathbf{Q}(\sqrt{d})$ in the case $d \equiv 1$ mod 4. In this case, $N(x-1-\sqrt{d}) = (x-1)^2 - d = x^2 - 2x + (1-d)$ has the coefficient of x divisible by 2 and the constant term divisible by 4, so $1 + \sqrt{d}$ is divisible by 2, that is, $\omega = (1+\sqrt{d})/2$ is integral over $\mathbf{Z}$. $F(x) = N(x - \omega) = x^2 - x + \frac{1-d}{4}$, $F'(x) = 2x - 1$, $F'(\omega) = 2\omega - 1 = \sqrt{d}$, $N[F'(\omega)] = [d]$, which shows that the method of §2.2 factors $[2]$ in $\mathbf{Q}(\sqrt{d}) = \mathbf{Q}(\omega)$ when $\alpha = \omega$. If $d \equiv 1 \bmod 8$, then $N(x - \omega) = x^2 - x + \frac{1-d}{4} \equiv x(x+1) \bmod 2$, so $[2] = [2, \omega][2, \omega + 1]$; if $d \equiv 5 \bmod 8$, then $N(x - \omega) \equiv x^2 + x + 1$ is irreducible mod 2, so $[2]$ is prime.

A cubic extension K of $\mathbf{Q}$ is one which can be obtained by adjoining a root of an irreducible polynomial $a_0 x^3 + a_1 x^2 + a_2 x + a_3$ with $a_i \in \mathbf{Z}$. If γ is a root of this equation in K, then $(3a_0\gamma)^3 + 3a_1(3a_0\gamma)^2 + 9a_0a_2(3a_0\gamma) + 27a_0^2 a_3 = 0$, and $\alpha = 3a_0\gamma + a_1$ satisfies $(\alpha - a_1)^3 + 3a_1(\alpha - a_1)^2 + 9a_0a_2(\alpha - a_1) + 27a_0^2 a_3 = 0$, which has the form $\alpha^3 + a\alpha + b = 0$, where $a, b \in \mathbf{Z}$. The norm of $F'(\alpha) = 3\alpha^2 + a$ is (use the basis 1, α, α^2 of K over $\mathbf{Q}$)

$$\begin{vmatrix} a & 0 & 3 \\ -3b & a - 3a & 0 \\ 0 & -3b & a - 3a \end{vmatrix} = 4a^3 + 27b^2,$$

so $[p]$ can be factored in K by factoring $x^3 + ax + b \bmod p$, *provided* $p^2 \nmid (4a^3 + 27b^2)$. As in the quadratic case, a cubic polynomial can be factored mod p by finding its roots, if any, mod p.

If $p^2 | (4a^3 + 27b^2)$, it may still be possible to factor $[p]$ in K by choosing a different generator of K over $\mathbf{Q}$, just as in the quadratic case $[2]$ was factored in $\mathbf{Q}(\sqrt{d})$ for $d \equiv 1 \bmod 4$ by writing $\mathbf{Q}(\sqrt{d}) = \mathbf{Q}(\omega)$, where $\omega = (1 + \sqrt{d})/2$. The classic case of a prime $[p]$ which can not be factored in an algebraic number field K by this method, no matter what generator of K over $\mathbf{Q}$ is used, is the case $p = 2$, $K = \mathbf{Q}(\alpha)$, where $\alpha^3 = \alpha^2 + 2\alpha + 8$ (Dedekind, Werke, vol. 1, p. 225).* With $\rho = 3\alpha - 1$, $(3\alpha)^3 = 3(3\alpha)^2 + 18(3\alpha) + 216$, $(\rho + 1)^3 = 3(\rho + 1)^2 +$

---

*See also Kronecker, II, p. 383.

$18(\rho+1)+216$, $\rho^3 = 21\rho+236$. Since $[4\cdot(-21)^3+27\cdot(-236)^2] = [2]^2[3]^3[-7^3+118^2] = [2]^2[3]^6[503]$, any $[p]$ other than $[2]$ or $[3]$ can be factored in K by factoring $x^3 - 21x - 236 \bmod p$. For example, $[5]$ can be factored by factoring $x^3 - 21x - 236 \equiv x^3 - x - 1 \equiv (x-2)(x^2+2x-2) \bmod 5$. The second factor is $(x+1)^2 - 3$, which is irreducible mod 5 because 3 is not a square mod 5. Therefore, $[5] = [5, \rho-2][5, \rho^2+2\rho-2]$ factors $[5]$ into divisors prime in K. (In terms of the original generator α, these divisors are easily found to be $[5, \alpha-1]$ and $[5, \alpha^2-2]$.)

Although the generator ρ does not give the factorization of $[3]$ $(x^3 - 21x - 236 \equiv x^3 + 1 \equiv (x+1)^3 \bmod 3)$, the generator α does, because $x^3 - x^2 - 2x - 8 \equiv x^3 - x^2 + x + 1 \bmod 3$ is irreducible mod 3 (it has no root mod 3). Therefore, $[3]$ is prime in K by Corollary (2) of §2.4.

Consider, finally, the factorization of $[2]$ in this K. If R_α does not absorb division by 2, then some polynomial of degree less than 3 in α with coefficients in $\mathbf{Z}$ is divisible by 2 without all its coefficients being divisible by 2. Since $\alpha^3 - \alpha^2 - 2\alpha - 8 = 0$, $\alpha^3 - \alpha^2 = \alpha^2(\alpha - 1)$ is divisible by 2, and a simple candidate for such a polynomial is $\alpha(\alpha - 1)$. Let $\beta = \alpha(\alpha - 1)/2$. Then $\beta \cdot 1 = -\frac{1}{2}\alpha + \frac{1}{2}\alpha^2$, $\beta \cdot \alpha = -\frac{1}{2}\alpha^2 + \frac{1}{2}(\alpha^2 + 2\alpha + 8) = 4 + \alpha$, and $\beta \cdot \alpha^2 = 4\alpha + \alpha^2$. Thus

$$N(x - \beta) = \begin{vmatrix} x & \frac{1}{2} & -\frac{1}{2} \\ -4 & x-1 & 0 \\ 0 & -4 & x-1 \end{vmatrix}$$

$$= \frac{1}{2}\begin{vmatrix} 2x & 1 & -1 \\ -4 & x-1 & 0 \\ 0 & -4 & x-1 \end{vmatrix} = \begin{vmatrix} x & 1 & -1 \\ -2 & x-1 & 0 \\ 0 & -4 & x-1 \end{vmatrix}$$

which shows that β is integral over $\mathbf{Z}$ and that R_α does *not* absorb division by 2. Since $\alpha \cdot 1 = \alpha$, $\alpha \cdot \alpha = \alpha^2$, and $\alpha \cdot \alpha^2 = 8 + 2\alpha + \alpha^2$,

$$N(x - \alpha u - \beta v) = \begin{vmatrix} x & -u + \frac{1}{2}v & -\frac{1}{2}v \\ -4v & x-v & -u \\ -8u & -2u-4v & x-u-v \end{vmatrix}$$

$$= \begin{vmatrix} x & -2u + v & -v \\ -2v & x - v & -u \\ -4u & -2u - 4v & x - u - v \end{vmatrix}.$$

Mod 2, the terms below the main diagonal are 0, so $N(x - \alpha u - \beta v) \equiv x(x - v)(x - u - v) \bmod 2$. If $R_{\alpha,\beta} = \mathbf{Z}[\alpha, \beta]$ absorbs division by 2, the factorization of [2] in K is therefore

$$[2] = [2, \alpha u + \beta v][2, \alpha u + \beta v - v][2, \alpha u + \beta v - u - v]$$

$$= [2, \alpha, \beta][2, \alpha, \beta + 1][2, \alpha + 1, \beta + 1]$$

That $R_{\alpha,\beta}$ does absorb division by 2—in fact, that $R_{\alpha,\beta}$ includes *all* elements of K integral over $\mathbf{Z}$—can be proved as follows. Since $2\beta = \alpha^2 - \alpha$ and since $1, \alpha, \alpha^2$ is a basis of K over $\mathbf{Q}$, the elements $1, \alpha, \beta$ are a basis of K over $\mathbf{Q}$. Therefore, any $\gamma \in K$ satisfies a relation $a\gamma = b + c\alpha + d\beta$, where $a, b, c, d \in \mathbf{Z}$ and $[a, b, c, d] = [1]$. By Corollary 1 of §1.30, $[a]^2$ divides disc$(1,\alpha,\beta)$. As was shown above, disc$(1,\rho,\rho^2) = [2]^2[3]^6[503]$. On the other hand,

$$1 = 1$$
$$\rho = -1 + 3\alpha$$
$$\rho^2 = 9\alpha^2 - 6\alpha + 1 = 9(2\beta + \alpha) - 6\alpha + 1$$
$$= 1 + 3\alpha + 18\beta.$$

By the theorem of §1.30, disc$(1, \rho, \rho^2) = [d]^2$ disc$(1,\alpha,\beta)$, where

$$d = \begin{vmatrix} 1 & 0 & 0 \\ -1 & 3 & 0 \\ 1 & 3 & 18 \end{vmatrix} = 2 \cdot 3^3,$$

so disc$(1,\alpha,\beta) = [503]$ and $[a]^2 |$ disc$(1, \alpha, \beta)$ implies $[a] = 1$, $a = \pm 1$. Thus, $\gamma \in R_{\alpha,\beta}$, as was to be shown.

That the method of §2.2 can not be used to factor [2] in this case can be seen as follows. Any $\gamma \in K$ integral over $\mathbf{Z}$ can be written in the form $\gamma = a + b\alpha + c\beta$. $P_1 = [2, \alpha, \beta]$ divides γ if a is even and $P_1|(\gamma - 1)$ if a is odd. Thus, $P_1|\gamma(\gamma - 1)$. Similarly, $P_2 = [2, \alpha + 1, \beta]$ and $P_3 = [2, \alpha, \beta + 1]$ both divide $\gamma(\gamma - 1)$. Since $[2] = P_1 P_2 P_3$ and the P_i are distinct and prime in K, it follows that $[2]|\gamma(\gamma - 1)$, that is, $\gamma^2 \equiv \gamma \bmod 2$. Thus, every polynomial $f(\gamma)$ in γ with coefficients in $\mathbf{Z}$ is congruent mod 2 to one of the four $0, 1, \gamma, \gamma + 1$. There are therefore at most 4 distinct divisors of the form $[2, f(\gamma)]$, two of which are $[2]$ and $[1]$. Thus, at least one of P_1, P_2, and P_3 is not of the form $[2, f(\gamma)]$.

§2.6 Integral Bases.

THEOREM. *Let K be an algebraic number field, and let γ_1, γ_2, ..., $\gamma_\mu \in K$ span K as a vector space over $\mathbf{Q}$. There is a basis β_1, β_2, ..., β_n of K over $\mathbf{Q}$ which spans the same $\mathbf{Z}$-module as the γ's, that is, for which $\beta_i = \sum b_{ij}\gamma_j$ and $\gamma_i = \sum c_{ij}\beta_j$, where the b_{ij} and c_{ij} are integers.*

PROOF: For each γ_i there is an integer a_i such that $a_i\gamma_i$ is integral over $\mathbf{Z}$. Therefore, there is an integer a such that $a\gamma_i$ is integral over $\mathbf{Z}$ for all i. If β_1, β_2, ..., β_n is a basis of K which generates the same $\mathbf{Z}$-module as the $a\gamma_i$, then $a^{-1}\beta_1$, $a^{-1}\beta_2$, ..., $a^{-1}\beta_n$ is a basis of K which generates the same $\mathbf{Z}$-module as the γ_i. Therefore, it will suffice to prove the theorem in the case where the γ_i are integral over $\mathbf{Z}$, which can be done using the following algorithm of Kronecker [**Kr1**, §7].

Since the γ_i span K, some subset of $n = [K : \mathbf{Q}]$ of the γ_i are a basis of K over $\mathbf{Q}$. The discriminant of a basis is $\neq [0]$, because if it were $[0]$, then, by the theorem of §1.30, the discriminant of *every* basis would be $[0]$, contrary to the proposition of §1.30 and (1) of §1.27. Therefore, some subset of n of the γ's has a nonzero discriminant. Moreover, if a subset of n of the γ's has nonzero discriminant then it is a basis of K over $\mathbf{Q}$, by the theorem of §1.30.

Step 1. Among the $\binom{\mu}{n}$ discriminants of subsets of n of the γ's, there is a smallest nonzero one. Therefore, the γ's can be rearranged, if necessary, to make $\mathrm{disc}(\gamma_1, \gamma_2, \ldots, \gamma_n) = [d]$, where d is a positive integer, and to make the discriminant of any subset of n of the γ's have the form $[d']$, where $d' = 0$ or $d' \geq d$.

Step 2. Since $\gamma_1, \gamma_2, \ldots, \gamma_n$ is a basis, every γ_i for $i > n$ (if any) can be written in the form $\gamma_i = \sum_{j=1}^{n} b_{ij}\gamma_j$ where $b_{ij} \in \mathbf{Q}$. By subtracting multiples of $\gamma_1, \gamma_2, \ldots, \gamma_n$ from γ_i, one can put all the coefficients b_{ij} in the range $0 \leq b_{ij} < 1$ without changing the $\mathbf{Z}$-module spanned by the γ's.

If, after step 2, the coefficients b_{ij} are all reduced to 0, then $\gamma_1, \gamma_2, \ldots, \gamma_n$ span the same $\mathbf{Z}$-module as the original γ's and the algorithm terminates. Otherwise, return to step 1. At step 1, d must be *decreased*, since if, for example, $b_{n+1,1} \neq 0$, then $\mathrm{disc}(\gamma_{n+1}, \gamma_2, \gamma_3, \ldots, \gamma_n) = [b_{n+1,1}]^2 \mathrm{disc}(\gamma_1, \gamma_2, \ldots, \gamma_n)$ by the theorem of §1.30, and $0 < b_{n+1,1} < 1$. Since d can only be decreased a finite number of times, the algorithm must terminate.

COROLLARY. *There is a basis* $\omega_1, \omega_2, \ldots, \omega_n$ *of* K *over* $\mathbf{Q}$ *with the property that an element* β *of* K *is integral over* $\mathbf{Z}$ *if and only if its representation* $\beta = \sum b_i \omega_i$ *in the basis has integer coefficients* b_i.

DEDUCTION: Apply the theorem to a set $\gamma_1, \gamma_2, \ldots, \gamma_\mu$ as in the lemma of §2.1.

Such a basis of K over $\mathbf{Q}$ is called an *integral basis* of K.

§2.7 Proof of the Theorem of §2.4.

Let the notation be as in §2.4. As was seen in §1.32, $V_{[p]}$, the ring of values of K at $[p]$ (see §1.23), is a vector space over k_p, the ring of values of $\mathbf{Q}$ at $[p]$. The field k_p is the field of quotients of the integral domain $\mathbf{Z}$ mod p, which is $\mathbf{Z}$ mod p itself, that is, the field $\mathbf{F}_p$ with p elements.

LEMMA 1. *The image in* $V_{[p]}$ *of an integral basis of* K *is a basis of* $V_{[p]}$ *over* $\mathbf{F}_p$. *In particular,* K *is p-regular* (§1.32).

PROOF: As was seen in the proof of Proposition 2, §1.32, $[V_{[p]}\colon \mathbf{F}_p] \le [K\colon\mathbf{Q}]$, so the lemma will be proved if it is shown that $\iota_p(\omega_1)$, $\iota_p(\omega_2)$, $\ldots$, $\iota_p(\omega_n)$ are linearly independent over $\mathbf{F}_p$, where ι_p is the natural map from elements of K finite at $[p]$ to $V_{[p]}$, and $\omega_1, \omega_2, \ldots, \omega_n$ is an integral basis of K. Let $a_1, a_2,$ $\ldots, a_n \in \mathbf{F}_p$ be such that $a_1\iota_p(\omega_1)+a_2\iota_p(\omega_2)+\cdots+a_n\iota_p(\omega_n) = 0$. Then $\gamma = b_1\omega_1+b_2\omega_2+\cdots+b_n\omega_n$ is integral over $\mathbf{Z}$ and satisfies $\iota_p(\gamma) = 0$, when $b_1, b_2, \ldots, b_n \in \mathbf{Z}$ satisfy $b_i \equiv a_i \bmod p$. Therefore, $\gamma = p\delta$ where δ is integral over $\mathbf{Z}$. Because the ω's are an integral basis, $\delta = c_1\omega_1 + c_2\omega_2 + \cdots + c_n\omega_n$ where $c_i \in \mathbf{Z}$. Because $\gamma = p\delta$, $b_i = pc_i$, so $b_i \equiv 0 \bmod p$. Since $a_i \equiv b_i \bmod p$, $a_i = 0$ for all i, so the $\iota_p(\omega_i)$ are linearly independent, as was to be shown.

Let $[p] = P_1^{e_1} P_2^{e_2} \cdots P_\mu^{e_\mu}$ be the factorization of $[p]$ as a product of powers of divisors prime in K. Lemma 1 combines with Proposition 1 of §1.32 to give

$$F(x) \equiv \prod_{i=1}^{\mu} g_i(x)^{e_i} \bmod p$$

where $F(x) = F(x, u_1, u_2, \ldots, u_\nu)$ is the homogeneous polynomial of degree $n = [K\colon\mathbf{Q}] = [V_{[p]}\colon\mathbf{F}_p]$ which is the determinant of the matrix representing the action of multiplication by $x - \bar\alpha$ on $V_{[p]}$ relative to a basis of $V_{[p]}$ over $\mathbf{F}_p$, and $g_i(x) = g_i(x, u_1, u_2, \ldots, u_\nu)$ is the homogeneous polynomial of degree $[V_{P_i}\colon\mathbf{F}_p]$ which is the determinant of the matrix representing the action of multiplication by $x - \bar\alpha$ on V_{P_i} relative to a basis of V_{P_i} over $\mathbf{F}_p$. Since $\mathbf{F}_p$ is a natural ring which is its own field of quotients, and since $V_{P_i} \supset \mathbf{F}_p$ is a field extension (§1.24, (4)) of finite degree ($[V_{P_i}\colon\mathbf{F}_p] \le [V_{[p]}\colon\mathbf{F}_p] = n$), g_i is simply the *norm* of $x - \bar\alpha$ when it is regarded as a polynomial with coefficients in the field extension $V_{P_i} \supset \mathbf{F}_p$ (§1.17).

LEMMA 2. *Let P be a divisor of $[p]$ prime in K, let ι_P denote the natural map from elements of K finite at P to the field*

$V_P \supset \mathbf{F}_p$, let $g(x)$ be the norm of $\iota_P(x - \bar{\alpha})$, and let R_α absorb division by p. Then g is irreducible over $\mathbf{F}_p$ and $P = [p, g(\bar{\alpha})]$.

(As above, $[p, g(\bar{\alpha})]$ means $[p, \hat{g}(\bar{\alpha})]$ where $\hat{g}$ is a polynomial with coefficients in $\mathbf{Z}$ congruent to $g \bmod p$.)

PROOF: Let $\delta \in K$ be integral over $\mathbf{Z}$. Since $\alpha_1, \alpha_2, \ldots, \alpha_\nu$ generate K over $\mathbf{Q}$ by assumption, $\delta = \phi(\alpha)/b$, where $\phi(\alpha) \in R_\alpha$ is a polynomial in the α's with coefficients in $\mathbf{Z}$, and $b \in \mathbf{Z}$ is nonzero. If $p|b$, say $b = pb'$, then $\phi(\alpha)/p = b'\delta$ is integral over $\mathbf{Z}$, and, by the assumption that R_α absorbs division by p, $\phi(\alpha)/p = \psi(\alpha) \in R_\alpha$. Thus $\delta = p\psi(\alpha)/b = \psi(\alpha)/b'$, where b' is divisible one less time by p than b is. Since this process can be repeated, if necessary, there is no loss of generality in assuming that $p \nmid b$. Then $\iota_P(\delta) = \iota_P(\phi(\alpha))/\iota_P(b)$, which shows that the element of V_P represented by δ is in the subfield of V_P generated over $\mathbf{F}_p$ by $\iota_P(\alpha_1), \iota_P(\alpha_2), \ldots, \iota_P(\alpha_\nu)$.

Any element of V_P can be represented by an element δ of K finite at P, say $[\delta] = B/C$, where B and C are integral divisors and $P \nmid C$. By Theorem 2, there is a $\delta_2 \in K$, integral over $\mathbf{Z}$, such that $[\delta_2] = CD$, where D is integral and $P \nmid D$. Then $\delta = \delta_1/\delta_2$, where δ_1 and δ_2 are integral over $\mathbf{Z}$ ($[\delta_1] = [\delta][\delta_2] = (B/C)CD = BD$) and δ_2 is not zero at P. Thus $\iota_P(\delta_1)$ and $\iota_P(\delta_2)$ are in the subfield of V_P generated over $\mathbf{F}_p$ by $\iota_P(\alpha_1)$, $\iota_P(\alpha_2), \ldots, \iota_P(\alpha_\nu)$. Moreover, $\iota_P(\delta_2) \neq 0$. Therefore $\iota_P(\delta) = \iota_P(\delta_1)/\iota_P(\delta_2)$ is in this subfield. Since $\iota_P(\delta)$ was arbitrary, $\iota_P(\alpha_1), \iota_P(\alpha_2), \ldots, \iota_P(\alpha_\nu)$ generate V_P over $\mathbf{F}_p$.

By the theory of finite fields, $V_P \supset \mathbf{F}_p$ is a *normal* extension. (In fact, the Galois group is cyclic of order $\kappa = [V_P : \mathbf{F}_p]$ and is generated by the automorphism $\zeta \mapsto \zeta^p$ of V_P over $\mathbf{F}_p$.) Therefore, the norm of a polynomial with coefficients in V_P is the product of its κ conjugates under the Galois group of $V_P \supset \mathbf{F}_p$. Thus $g = \prod_S (x - S\iota_P(\bar{\alpha}))$, where S ranges over the Galois group of V_P over $\mathbf{F}_p$. Let h be a polynomial with coefficients in $\mathbf{F}_p$ which is not relatively prime to g. Then, as a polynomial with coefficients in $V_P \supset \mathbf{F}_p$, h must be divisible by one of the factors $x - S\iota_P(\bar{\alpha})$ of g (unique factorization over V_P). Then $x - S'S\iota_P(\bar{\alpha})$ divides h for all S' in the Galois group of V_P over $\mathbf{F}_p$ (h has coefficients in $\mathbf{F}_p$). But these factors of h are *distinct* because $x - S'S\iota_P(\bar{\alpha}) = x - S''S\iota_P(\bar{\alpha})$ implies $S'S$ and $S''S$ have the same effect on the coefficients $\iota_P(\alpha_i)$ of $\iota_P(\bar{\alpha})$, and these coefficients generate V_P over $\mathbf{F}_p$, so $S'S$ and $S''S$ can háve the same effect on them only if $S'S = S''S$. Thus h must be divisible by all κ factors of g, that is, $g|h$, which shows that g is irreducible over $\mathbf{F}_p$.

Let Q be a divisor of $[p, g(\bar{\alpha})]$ which is prime in K. Then $Q|[p]$, and, as was just shown, $g_Q = N(x - \iota_Q(\bar{\alpha}))$ is irreducible over $\mathbf{F}_p$, where $\iota_Q(\bar{\alpha})$ is the polynomial with coefficients in V_Q represented by $\bar{\alpha}$. On the other hand, Q divides $\hat{g}(\bar{\alpha})$, which is the remainder when $\hat{g}(x)$ is divided by $x - \bar{\alpha}$; therefore, $x - \iota_Q(\bar{\alpha})$ divides $\iota_Q\hat{g} = g$ as a polynomial with coefficients in V_Q. It follows that $g_Q = N(x - \iota_Q(\bar{\alpha}))$ divides $N_{V_Q}g$, which is a power of g. Since g_Q and g are irreducible over $\mathbf{F}_p$ and monic in x, $g_Q = g$. In particular, $\deg g_Q = \deg g = \kappa$, which implies that $[V_Q : \mathbf{F}_p] = \kappa = [V_P : \mathbf{F}_p]$, and therefore that V_Q and V_P are isomorphic (they are both splitting fields of $X^{p^\kappa} - X$ over $\mathbf{F}_p$). Let $\theta : V_P \to V_Q$ be an isomorphism. Application of θ to the equation $g = \prod(x - S\iota_P(\bar{\alpha}))$ gives g on the left ($\theta(1) = 1$, so θ is the identity on $\mathbf{F}_p$) and gives $\prod(x - \theta S\iota_P(\bar{\alpha}))$ on the right. Since $x - \iota_Q(\bar{\alpha})$ divides $N(x - \iota_Q(\bar{\alpha})) = g$, it follows that $x - \iota_Q(\bar{\alpha}) = x - \theta S\iota_P(\bar{\alpha})$ for some automorphism S of V_P. Thus, for this S, $\iota_Q(\alpha_i) = \theta S\iota_P(\alpha_i)$ for all i. Therefore, if $\phi(\alpha) \in K$ can be represented as a polynomial in the α's with coefficients in $\mathbf{Z}$, $\iota_Q(\phi(\alpha)) = \theta S\iota_P(\phi(\alpha))$. In particular, $\iota_Q(\phi(\alpha)) = 0$ if and only if $\iota_P(\phi(\alpha)) = 0$, that is, $Q|\phi(\alpha)$ if and only if $P|\phi(\alpha)$. If Q were not equal to P, there would be, by Theorem 2, an element δ of K with $[\delta] = PC$, where C is integral and $Q \nmid C$. As was shown above, δ could be written in the form $\delta = \phi(\alpha)/b$, where $p \nmid b$, and $[\phi(\alpha)] = [\delta][b]$ would be divisible by P and not by Q, contrary to what was just shown. Therefore, Q must be P, which shows that $[p, g(\bar{\alpha})] = P^j$ for some $j \geq 0$. It remains to show that $j = 1$.

Since $\widehat{g}(\bar{\alpha})$ is the remainder when $\widehat{g}(x)$ is divided by $x - \bar{\alpha}$, $\iota_P \widehat{g}(\bar{\alpha})$ is the remainder when $\iota_P g(x) = \prod(x - S\iota_P(\bar{\alpha}))$ is divided by $x - \iota_P(\bar{\alpha})$, which is 0; that is, $P|\widehat{g}(\bar{\alpha})$. Therefore $j \geq 1$, and it remains only to show that $P^2 \nmid [p, g(\bar{\alpha})]$.

Suppose $P^2 \| [p]$. The lemma will be proved by showing that in this case $P^2 \nmid \widehat{g}(\bar{\alpha})$. Since $P^2 | p$, V_{P^2} is a vector space over $\mathbf{F}_p$. Since $[V_P : \mathbf{F}_p] = \kappa < \infty$, V_P is generated over $\mathbf{F}_p$ by a single element of V_P (§1.25), say by $\iota_P \zeta$, where $\zeta \in K$ is finite at P. Then the images of $1, \zeta, \zeta^2, \ldots, \zeta^{\kappa-1}$ in V_P are a basis of V_P over $\mathbf{F}_p$. Let $\eta \in K$ be integral over $\mathbf{Z}$ and divisible by P but not by P^2. As in the proof of Proposition 1, §1.32, the images of $1, \zeta, \zeta^2, \ldots, \zeta^{\kappa-1}, \eta, \eta\zeta, \ldots, \eta\zeta^{\kappa-1}$ in V_{P^2} are a basis of V_{P^2} as a vector space over $\mathbf{F}_p$. In order to have a convenient description of V_{P^2} as a *ring*, it is helpful to adjust the choice of ζ in the following way.

The norm of $x - \iota_P \zeta$ (a polynomial with coefficients in $V_P \supset \mathbf{F}_p$) is a monic polynomial $\Phi(x)$ of degree κ in x with coefficients in $\mathbf{F}_p$ of which $\iota_P \zeta$ is a root. Let ζ_0 be the element of V_{P^2} represented by ζ. Then $\Phi(\zeta_0)$ is a well-defined element of $V_{P^2} \supset \mathbf{F}_p$, say $\Phi(\zeta_0) = a_0 + a_1\zeta_0 + \cdots + a_{\kappa-1}\zeta_0^{\kappa-1} + b_0\eta_0 + b_1\eta_0\zeta_0 + \cdots + b_{\kappa-1}\eta_0\zeta_0^{\kappa-1}$, where η_0 is the element of V_{P^2} represented by η, and the a's and b's are in $\mathbf{F}_p$. The image of $\Phi(\zeta_0)$ under $V_{P^2} \to V_P$ is $\Phi(\iota_P \zeta) = 0$ on the one hand and $a_0 + a_1\iota_P\zeta + \cdots + a_{\kappa-1}\iota_P\zeta^{\kappa-1}$ on the other, which implies the a_i are all zero, that is, $\Phi(\zeta_0) = \eta_0(b_0 + b_1\zeta_0 + \cdots + b_{\kappa-1}\zeta_0^{\kappa-1})$. For any $\delta \in V_{P^2}$, since $\eta_0^2 = 0$, $\Phi(\zeta_0 + \eta_0\delta) = \Phi(\zeta_0) + \eta_0\delta\Phi'(\zeta_0) = \eta_0(b_0 + b_1\zeta_0 + \cdots + b_{\kappa-1}\zeta_0^{\kappa-1} + \delta\Phi'(\zeta_0))$. The image of $\Phi'(\zeta_0)$ in V_P is $\Phi'(\iota_P\zeta) \neq 0$ (since Φ is irreducible over $\mathbf{F}_p$, it has no multiple roots in $V_P \supset \mathbf{F}_p$). Since V_P is a field, there is a γ in K finite at P such that $\iota_P(\gamma)\Phi'(\iota_P\zeta) = 1$. Let $\zeta_1 = \zeta_0 - \eta_0\gamma_0 \sum_{i=0}^{\kappa-1} b_i\zeta_0^i$ where γ_0 is the element of V_{P^2} represented by γ. Then $\Phi(\zeta_1) = \eta_0\left(\sum b_i\zeta_0^i - \gamma_0 \sum b_i\zeta_0^i\Phi'(\zeta_0)\right) = \eta_0\left(1 - \gamma_0\Phi'(\zeta_0)\right)\sum b_i\zeta_0^i$. Since the image of $1 - \gamma_0\Phi'(\zeta_0)$ in V_P is $1 - \iota_P(\gamma)\Phi'(\iota_P\zeta) = 0$, its expansion in the basis $\zeta_0^i\eta_0^j$

$(0 \leq i < \kappa, 0 \leq j < 2)$ of V_{P^2} over $\mathbf{F}_p$ contains no terms with $j = 0$; therefore, since $\eta_0^2 = 0$, $\Phi(\zeta_1) = 0$. Since the image of ζ_1 in V_P is $\iota_P \zeta$, which is a generator of V_P over $\mathbf{F}_p$, 1, ζ_1, ζ_1^2, $\ldots$, $\zeta_1^{\kappa-1}$, η_0, $\eta_0\zeta_1$, $\ldots$, $\eta_0\zeta_1^{\kappa-1}$ is a basis of V_{P^2} over $\mathbf{F}_p$. Multiplication in V_{P^2} relative to this basis is described by the two relations $\Phi(\zeta_1) = 0$ (which gives ζ_1^κ as a linear combination of lower powers of ζ_1) and $\eta_0^2 = 0$.

Let R denote the subring $\mathbf{F}_p[\zeta_1]$ of V_{P^2} generated by 1 and ζ_1. An element of V_{P^2} can be written in one and only one way in the form $s + \eta_0 t$, where s, $t \in R$. Let $\iota_{P^2}\alpha_i = s_i + \eta_0 t_i$ for $i = 1, 2, \ldots, \nu$ and let $\bar{s} = \sum s_i u_i$ and $\bar{t} = \sum t_i u_i$, where the u_i are indeterminates as before. Then $\iota_{P^2}\bar{\alpha} = \bar{s} + \eta_0\bar{t}$ and $g(\iota_{P^2}\bar{\alpha}) = g(\bar{s} + \eta_0\bar{t}) = g(\bar{s}) + \eta_0\bar{t}g'(\bar{s})$ (because $\eta_0^2 = 0$), where $g(x) = g(x, u_1, u_2, \ldots, u_\nu)$ is the norm of $x - \iota_P\bar{\alpha}$, a polynomial with coefficients in $\mathbf{F}_p \subset V_{P^2}$, where $g'(x)$ is its derivative with respect to x, and where $g(\bar{s})$ and $g'(\bar{s})$ are the polynomials with coefficients in R obtained by setting $\bar{s}$ in place of x in these polynomials. The image of $g(\bar{s})$ in V_P is $g(\iota_P\bar{\alpha})$ (because $\iota_{P^2}\bar{\alpha} = \bar{s} + \eta_0\bar{t}$ has the same image in V_P as $\iota_P\bar{\alpha}$ and $\bar{s}$) which is 0 (by the definition of g). Therefore, since $g(\bar{s})$ is in R, $g(\iota_{P^2}\bar{\alpha}) = \eta_0\bar{t}g'(\bar{s})$. If $\bar{t}$ were 0, $\iota_{P^2}\alpha_i$ would be $s_i \in R$ for all i, and $\iota_{P^2}\phi(\alpha)$ would be in R for any $\phi(\alpha) \in R_\alpha$; but η is integral over $\mathbf{Z}$, so $\eta = \phi(\alpha)/b$ where $\phi(\alpha) \in R_\alpha$, and where $b \in \mathbf{Z}$ is not divisible by p, which gives $\eta_0 = \iota_{P^2}\eta = \iota_{P^2}(bc\eta) = \iota_{P^2}(c\phi(\alpha)) = \iota_{P^2}(c)\iota_{P^2}(\phi(\alpha))$, where $c \in \mathbf{Z}$ satisfies $bc \equiv 1 \bmod p$. Since $\eta_0 \notin R$, $\iota_{P^2}(c) \in R$, and R is a field (isomorphic to V_P under $V_{P^2} \to V_P$), this shows that $\iota_{P^2}(\phi(\alpha)) \notin R$ and therefore that $\bar{t} \neq 0$. The image of $g'(\bar{s})$ in V_P is $g'(\iota_P\bar{\alpha})$, which is not zero, because, as was shown above, $x - \iota_P\bar{\alpha}$ is a simple factor of $g(x)$. Therefore $g'(\bar{s}) \neq 0$. Since R is a field, $\bar{t}g'(\bar{s})$ is a polynomial with coefficients in R which is not 0, so $\eta_0\bar{t}g'(\bar{s}) = g(\iota_{P^2}\bar{\alpha})$ is a polynomial with coefficients in V_{P^2} which is not 0. Thus $x - \bar{\alpha}$ does not divide $\hat{g}(x)$ when both are regarded as polynomials with coefficients in $V_{P^2} \supset \mathbf{F}_p$. In other words, P^2 does not divide $\hat{g}(\bar{\alpha})$, as was to be shown.

PROOF OF THE THEOREM: As was shown above, $F(x) \equiv \prod_{i=1}^{\mu} g_i(x)^{e_i} \bmod p$. Since the $g_i(x)$ are irreducible over $\mathbf{F}_p$ by Lemma 2, and since they are monic in x, they can be found by factoring $F(x) \bmod p$. Then the factorization $[p] = \prod P_i^{e_i}$ can be found using the formula $P_i = [p, g_i(\bar{\alpha})]$ of Lemma 2, which is the assertion of the theorem.

§2.8 Dedekind's Discriminant Theorem.

As the lemma of §2.1 shows, there exist elements α_1, α_2, ..., $\alpha_\nu \in K$ integral over $\mathbf{Z}$ such that $R_\alpha = \mathbf{Z}[\alpha_1, \alpha_2, \ldots, \alpha_\nu]$ includes *all* elements of K integral over $\mathbf{Z}$. The discriminant and the different of this ring are called the *discriminant of the field K* and the *different of the field K*, denoted disc(K) and dif(K), respectively.

THEOREM. (Dedekind) *Let K be an algebraic number field. A divisor P of $[p]$ prime in K which divides $[p]$ exactly e times divides* dif(K) *at least $e - 1$ times. It divides* dif(K) *exactly $e - 1$ times except possibly when $p|e$.*

PROOF: Since dif(K) $= [F'(\bar{\alpha})]$, where $F(x) = N(x - \bar{\alpha})$ and α_1, α_2, ..., α_ν are such that R_α includes all elements of K integral over $\mathbf{Z}$, the theorem deals with the number of times a prime divisor P of $[p]$ divides $F'(\bar{\alpha})$. As was shown in §2.7, $F(x) \equiv \prod_{i=1}^{\mu} g_i(x)^{e_i} \bmod p$, where $[p] = \prod_i P_i^{e_i}$ is the prime factorization of $[p]$ in K and $g_i(x)$ is the norm of $x - \bar{\alpha}$ regarded as a polynomial with coefficients in the field $V_{P_i} \supset \mathbf{F}_p$ (R_α absorbs division by p). Therefore, $F'(x) \equiv \sum_{i=1}^{\mu} e_i g_i'(x) g_i(x)^{-1} F(x) \bmod p$. Therefore $F'(\bar{\alpha}) \equiv \sum_{i=1}^{\mu} e_i g_i'(\bar{\alpha}) g_i(\bar{\alpha})^{-1} F(\bar{\alpha}) \bmod P_1^{e_1}$. Since $P_1|g_1(\bar{\alpha})$, $P_1^{e_1}$ divides all terms on the right except the first, and $P_1^{e_1-1}$ divides the first. Therefore, $P_1^{e_1-1}|F'(\bar{\alpha})$. It remains to show that, if $p \nmid e_1$, then $P_1^{e_1}$ does not divide $F'(\bar{\alpha})$, or, what is the same, that $P_1^{e_1}$ does not divide $e_1 g_1'(\bar{\alpha}) g_1(\bar{\alpha})^{e_1-1} \prod_{i=2}^{\mu} g_i(\bar{\alpha})^{e_i}$. Since P_1 is prime, the multiplicity with which it divides a product is the sum of the multiplicities with which it divides the factors. Since $P_i = [p, g_i(\bar{\alpha})]$ is not divisible by P_1, $\prod_{i=2}^{\mu} g_i(\bar{\alpha})^{e_i}$ is not divisible by P_1. Since $N(P_1)$ does not divide $N(e_1)$ (p does not divide e_1), P_1 does not divide e_1. Since $x - \iota_{P_1}\bar{\alpha}$ is a simple factor of $g_1(x)$ over V_{P_1} (§2.7), $g_1'(\bar{\alpha})$ is not divisible by P_1. If $e_1 = 1$, it follows that $P_1 \nmid e_1 g_1'(\bar{\alpha}) g_1(\bar{\alpha})^{e_1 - 1} \prod_{i>1} g_i(\bar{\alpha})^{e_i} = e_1 g_1'(\bar{\alpha}) \prod_{i>1} g_i(\bar{\alpha})^{e_i}$, as desired. If $e_1 > 1$, then $P_1^2 | p$ and the equation $P_1 = [p, g_1(\bar{\alpha})]$ implies P_1 divides $g_1(\bar{\alpha})$ exactly once and therefore divides $g_1(\bar{\alpha})^{e_1 - 1}$ exactly $e_1 - 1$ times, and the theorem is proved.

COROLLARY. *A prime $p \in \mathbf{Z}$ ramifies in K if and only if $[p, \mathrm{dif}(K)] \neq [1]$.*

DEDUCTION: If p ramifies in K, then some divisor P of $[p]$ prime in K divides $[p]$ exactly e times where $e > 1$. By the theorem, P^{e-1} divides both $[p]$ and $\mathrm{dif}(K)$, so $[p, \mathrm{dif}(K)] \neq [1]$. Conversely, if $[p, \mathrm{dif}(K)] \neq [1]$ then some divisor P of $[p]$ prime in K also divides $\mathrm{dif}(K)$. Let e be the multiplicity with which P divides p. If $p | e$, then $e > 1$, so p ramifies in K. If $p \nmid e$, then, by the theorem, $e - 1 > 0$, so p ramifies in K.

§2.9 Differents and Discriminants.

THEOREM (Dedekind*). *For algebraic number fields K, the discriminant is the norm of the different, $\mathrm{disc}(K) = N(\mathrm{dif}(K))$.*

PROOF: Let $\alpha_1, \alpha_2, \ldots, \alpha_n$ be an integral basis of K. Then R_α is the ring of all elements of K integral over $\mathbf{Z}$ and $\mathrm{dif}(K) = \mathrm{dif}_K(R_\alpha)$, $\mathrm{disc}(K) = \mathrm{disc}_K(R_\alpha)$. Since $\alpha_1, \alpha_2, \ldots, \alpha_n$ span R_α over $\mathbf{Z}$, $\mathrm{disc}_K(R_\alpha)$ is by definition $[\det(USU^t)]$, where U is an $n \times n$ matrix of indeterminates and S is the $n \times n$ symmetric

---

*$N(\mathrm{dif}(K)) = \prod_{i<j} [\bar{\alpha}^{(j)} - \bar{\alpha}^{(i)}]^2$ is in essence what Kronecker called "the discriminant of the fundamental equation". Kronecker conjectured that it was equal to $\mathrm{disc}(K)$, and he proved partial results in this direction [Kr1, §25], but he admitted he could not prove it for general r. Probably he could prove the case $r = \mathbf{Z}$ stated here, but he never published a proof.

matrix of integers $(\operatorname{tr}(\alpha_i\alpha_j))$. Since $[\det(U)] = [1]$, $\operatorname{disc}(K) = [\det(S)]$.

On the other hand, as was seen in §1.27, $\operatorname{dif}_K(R_\alpha) = \prod[\bar{\alpha} - \bar{\alpha}^{(i)}]$, where the product is over all $n - 1$ conjugates $\bar{\alpha}^{(i)}$ of $\bar{\alpha}$, other than $\bar{\alpha}$ itself, in a normal extension of $\mathbf{Q}$ containing K. Therefore, $N\left(\operatorname{dif}_K(R_\alpha)\right) = \prod_{i<j}[\bar{\alpha}^{(j)} - \bar{\alpha}^{(i)}]^2$ (Proposition 2, §1.25—see also the proposition of §1.30). This divisor is the divisor represented by the square of the determinant of the matrix A whose columns are the n conjugates of 1, $\bar{\alpha}$, $\bar{\alpha}^2$, $\ldots$, $\bar{\alpha}^{n-1}$. Thus $N\left(\operatorname{dif}(K)\right) = [\det(AA^t)]$ for this A. Now each $\bar{\alpha}^i$ is of the form $\phi_{i1}\alpha_1 + \phi_{i2}\alpha_2 + \cdots + \phi_{in}\alpha_n$, where the ϕ_{ij} are polynomials in u_1, u_2, $\ldots$, u_n (homogeneous of degree i) with coefficients in $\mathbf{Z}$. Thus, $A = \Phi B$, where Φ is the $n \times n$ matrix (ϕ_{ij}) and B is the matrix whose columns are the n conjugates of α_1, α_2, $\ldots$, α_n. The entry in the ith row and the jth column of BB^t is $\sum_{\rho=1}^n \alpha_i^{(\rho)}\alpha_j^{(\rho)} = \operatorname{tr}(\alpha_i\alpha_j)$, that is, $BB^t = S$. Thus, $N\left(\operatorname{dif}(K)\right) = [\det(\Phi BB^t\Phi^t)] = [\det(S)(\det\Phi)^2] = \operatorname{disc}(K)[\det\Phi]^2$. It is to be shown that $[\det\Phi] = [1]$, that is, $\det\Phi$ is primitive.

If $\det\Phi$ were not primitive, there would be a prime p such that $\det\Phi \equiv 0 \bmod p$. There would then be a row matrix $v = (v_1(u), v_2(u), \ldots, v_n(u))$ of polynomials in u_1, u_2, $\ldots$, u_n with coefficients in $\mathbf{Z}$, such that $v\Phi \equiv 0 \bmod p$, but $v \not\equiv 0 \bmod p$. Then v times the column matrix 1, $\bar{\alpha}$, $\bar{\alpha}^2$, $\ldots$, $\bar{\alpha}^{n-1}$ would be $v\Phi$ times the column matrix α_1, α_2, $\ldots$, α_n, so $v_1(u) + v_2(u)\bar{\alpha} + \cdots + v_n(u)\bar{\alpha}^{n-1} \equiv 0 \bmod p$. Let $G(x) = v_1(u) + v_2(u)x + \cdots + v_n(u)x^{n-1}$. Then $G(x) \not\equiv 0 \bmod p$ but $G(\bar{\alpha})$ would be $0 \bmod p$. That this is impossible, and therefore that $\det\Phi$ is primitive, follows from:

LEMMA (Hensel). *Let $G(x)$ be a polynomial in x, u_1, u_2, $\ldots$, u_n with coefficients in $\mathbf{Z}$ such that $G(\bar{\alpha}) \equiv 0 \bmod p$. Then $G(x)$ is divisible by $F(x) = N(x - \bar{\alpha}) \bmod p$. In particular, if $G \not\equiv 0 \bmod p$ then $\deg_x G \geq \deg_x F = n$.*

PROOF: Let $G \equiv h_1^{\epsilon_1} h_2^{\epsilon_2} \cdots h_\sigma^{\epsilon_\sigma} \bmod p$ be the decomposition of

G as a product of powers of distinct, irreducible polynomials with coefficients in $\mathbf{F}_p$, monic in x. Since $G(\bar{\alpha}) \equiv 0 \bmod P_1$ and P_1 is prime, $h_i(\bar{\alpha}) \equiv 0 \bmod P_1$ for at least one i. As a polynomial with coefficients in V_{P_1}, $h_i(x)$ therefore has the common factor $x - \iota_{P_1}\bar{\alpha}$ with $g_1 = N(x - \iota_{P_1}\bar{\alpha})$. Then, since $h_i(x)$ has coefficients in $\mathbf{F}_p$, it is divisible by all $\deg g_1$ distinct linear factors of g_1, so $h_i(x)$ is divisible by $g_1(x)$. Since $h_i(x)$ is irreducible over $\mathbf{F}_p$ and monic in x, $h_i(x) = g_1(x)$. Thus, exactly one of the $h_i(x)$ is equal to $g_1(x)$, and $P_1 \nmid h_j(\bar{\alpha})$ for $j \neq i$. Since $P_1^{e_1}$ divides $[p]$ and $[p]$ divides $G(\bar{\alpha})$, and since P_1 divides $h_i(\bar{\alpha})$ once and $h_j(\bar{\alpha})$ not at all for $j \neq i$, $\epsilon_i \geq e_1$, that is, $g_1^{e_1}$ divides $G(x)$. Since the same is true of all prime power factors of $F(x) = g_1(x)^{e_1} g_2(x)^{e_2} \cdots g_\mu(x)^{e_\mu}$, it follows that $F(x) | G(x) \bmod p$, as was to be shown.

COROLLARY. *A prime p ramifies in K if and only if* $[p] | \operatorname{disc}(K)$.

DEDUCTION: See the Corollary of §2.8.

§2.10 Cyclotomic Fields.

For positive integers n, let Λ_n be the splitting field of the polynomial $x^n - 1$ over $\mathbf{Q}$. These fields are called *cyclotomic* fields. A cyclotomic field Λ_n contains a *primitive* nth root of unity—that is, an element α which satisfies $\alpha^n = 1$ but does not satisfy $\alpha^m = 1$ for $0 < m < n$—and Λ_n is generated over $\mathbf{Q}$ by any such α. An element of the Galois group of Λ_n over $\mathbf{Q}$ obviously carries a primitive nth root of unity α to another primitive nth root of unity. Since the powers of a primitive nth root of unity $\alpha, \alpha^2, \ldots, \alpha^n = 1$ are n distinct roots of $x^n - 1$, every nth root of unity in Λ_n is a power of α. Therefore, an element of the Galois group of $\Lambda_n = \mathbf{Q}(\alpha)$ over $\mathbf{Q}$ carries α to α^j for some integer j, and j determines the automorphism. Since the composition of $\alpha \mapsto \alpha^j$ and $\alpha \mapsto \alpha^i$ is $\alpha \mapsto \alpha^{ij}$, and since α^j depends only on the class of $j \bmod n$, the Galois group of Λ_n over $\mathbf{Q}$ is isomorphic in this way to a subgroup of the multiplicative group of invertible elements in the ring of integers mod n. By a basic theorem (see, for example, [**E3**, §70]) the Galois group of Λ_n over **Q** is the *entire* group of invertible elements of the ring of integers mod n.

Since the Galois group of Λ_n over **Q** is commutative, every subfield K of Λ_n is normal, and there is a natural homomorphism—namely, the restriction mapping—from the Galois group of Λ_n over **Q** to the Galois group of K over **Q**.

THEOREM. *Let K be a subfield of a cyclotomic field Λ_n. For any prime integer p which does not divide n, $[p]$ is a product of distinct divisors $P_1 P_2 \cdots P_\mu$ prime in K. An element of the Galois group of K over **Q** leaves P_1 fixed (and therefore leaves P_i fixed for all i) if and only if it is a power of the restriction to K of the automorphism $\alpha \mapsto \alpha^p$ of Λ_n (α a primitive nth root of unity).*

PROOF: Let α be a primitive nth root of unity in Λ_n and let $\Phi_n(x) = N_{\Lambda_n}(x - \alpha)$. Since $\alpha^n = 1$, $x^n - 1 = \Phi_n(x)Q(x)$, where Q is a polynomial with coefficients in **Z**. Then $nx^{n-1} = \Phi'_n(x)Q(x) + \Phi_n(x)Q'(x)$, from which it follows that $n\alpha^{n-1} = \Phi'_n(\alpha)Q(\alpha)$, $n = \Phi'_n(\alpha)\alpha Q(\alpha)$. Therefore, $\Phi'_n(\alpha)$ divides n and $N[\Phi'_n(\alpha)]$ divides $N[n]$, which is a power of $[n]$. If p ramified in K, then it would ramify in Λ_n, and $[p, \Phi'_n(\alpha)]$ would not be $[1]$ (§2.8). The norm of $[p, \Phi'_n(\alpha)]$ would therefore be of the form $[p]^j$ for $j > 0$, and it would divide a power of $[n]$, contrary to the assumption $p \nmid n$. Thus, $[p] = P_1 P_2 \cdots P_\mu$, where the P_i are the distinct conjugates under the Galois group of K over **Q** of any one divisor of $[p]$ prime in K (§1.22).

Let ξ be an element of the Galois group of K over **Q** for which $\xi(P_1) = P_1$. Then there is an element ξ' of the Galois group of Λ_n over **Q** whose restriction to K is ξ. (The restriction map is a homomorphism from a group with $[\Lambda_n : \mathbf{Q}] = [\Lambda_n : K][K : \mathbf{Q}]$ elements to a group with $[K : \mathbf{Q}]$ elements whose kernel has $[\Lambda_n : K]$ elements. Therefore it is onto.) Let $P_1 = P'_1 P'_2 \cdots P'_\sigma$ be the factorization of P_1 into divisors prime in Λ_n. By the proposition of §1.22, the Galois group of Λ_n over K acts transitively on the factors P_i' of P_1. Since $\xi'(P_1) = P_1$, $\xi'(P_1') = P_i'$ for some $i = 1, 2, \ldots, \sigma$, so there is a θ in the Galois group of Λ_n over K such that $\theta(P_1') = \xi'(P_1')$. Then the element $\psi = \theta^{-1}\xi'$ of the Galois group of Λ_n over $\mathbf{Q}$ satisfies $\psi(P_1') = P_1'$ and its restriction to K is the given element ξ of the Galois group of K over $\mathbf{Q}$. It will suffice to show that an element ψ of the Galois group of Λ_n over $\mathbf{Q}$ leaves fixed a divisor P' of $[p]$ prime in Λ_n if and only if it is a power of $\alpha \mapsto \alpha^p$.

If $\psi(P') = P'$, where P' is prime in Λ_n, then ψ induces an automorphism of the field $V_{P'}$ (the ring of values of Λ_n at P'). An automorphism of the finite field $V_{P'}$ is necessarily a power of the automorphism of raising to the power p. Therefore, $\hat{\psi}\big(\iota_{P'}(\alpha)\big) = \iota_{P'}(\alpha)^{p^i}$, for some positive integer i, where $\iota_{P'}(\alpha)$ is the element of $V_{P'}$ represented by α and $\hat{\psi}$ denotes the automorphism of $V_{P'}$ induced by ψ. By definition, $\hat{\psi}\big(\iota_{P'}(\alpha)\big) = \iota_{P'}\big(\psi(\alpha)\big)$. Because $\iota_{P'}$ is a homomorphism, $\iota_{P'}(\alpha)^{p^i} = \iota_{P'}(\alpha^{p^i})$. Therefore, $\iota_{P'}\big(\psi(\alpha)\big) = \iota_{P'}(\alpha^{p^i})$, which implies $\iota_{P'}\big(\psi(\alpha) - \alpha^{p^i}\big) = 0$, which is to say $P'|\big(\psi(\alpha) - \alpha^{p^i}\big)$. Since $\psi(\alpha) = \alpha^j$ for some integer j, $P'|(\alpha^j - \alpha^{p^i})$. If j were not congruent to p^i mod n, then P' would divide $\alpha^j - \alpha^k$, where α^j and a^k were distinct roots of $\Phi_n(x)$; then some conjugate of P' would divide $\alpha - \alpha^s$, where $\alpha^s \neq \alpha$ was a root of $\Phi_n(x)$, and consequently would divide $\Phi_n'(\alpha) = \prod_t(\alpha - \alpha^t)$ where α^t ranges over the roots of $\Phi_n(x)$ other than α, which in turn divides n, contrary to $[p, n] = [1]$. Therefore, $j \equiv p^i \bmod n$ and ψ has the same effect on α as the ith power of $\alpha \mapsto \alpha^p$. Since α generates Λ_n over $\mathbf{Q}$, ψ is therefore the ith power of $\alpha \mapsto \alpha^p$.

It remains only to show that the element, call it η, of the Galois group of Λ_n over $\mathbf{Q}$ which carries α to α^p leaves fixed all divisors P' of $[p]$ prime in Λ_n (which implies, of course, that its restriction to K leaves fixed all divisors P of $[p]$ prime in

K). Let P' be written in the form $P' = [p, \beta]$ (Corollary (1), §1.20). Since β is integral over $\mathbf{Z}$ and since R_α absorbs division by p, $b\beta = \phi(\alpha) \in R_\alpha$, where $b \in \mathbf{Z}$ is not divisible by p. By Fermat's theorem, $\phi(\alpha)^p - \phi(\alpha^p)$ is of the form $p \cdot h(\alpha)$, where $h(\alpha) \in R_\alpha$. Therefore, P' divides $\phi(\alpha)^p - \phi(\alpha^p) = (b\beta)^p - \eta(b\beta) = b^p\beta^p - b\eta(\beta)$. Since $P'|\beta$, it follows that $P'|b\eta(\beta)$, and, since P' is prime and does not divide b, that $P'|\eta(\beta)$. Thus, P' divides $[p, \eta(\beta)] = \eta(P')$. Since $\eta(P')$ is prime in Λ_n and therefore irreducible in Λ_n, $P' = \eta(P')$, as was to be shown.

§2.11.

COROLLARY 1. *Let K be a subfield of the cyclotomic field Λ_n. The form of the factorization $[p] = (P_1 P_2 \cdots P_\mu)^\nu$ of $[p]$ in K depends only on the class of p mod n. If $p \nmid n$, then $\nu = 1$ and $\mu = [K : \mathbf{Q}]/\lambda$, where λ is the least positive integer for which $\alpha \mapsto \alpha^{p^\lambda}$ is the identity on K.*

DEDUCTION: If $p|n$ then p is the only prime in its class mod n, and there is nothing to prove. If $p \nmid n$, then, by the theorem, $[p]$ is a product of distinct prime factors, and the number μ of these factors describes the form of the factorization of $[p]$. Since λ depends only on the class of p mod n, the second statement of the corollary implies the first. By the proposition of §1.22, the Galois group of K acts transitively on the P_i. Since this group has $[K : \mathbf{Q}]$ elements, the second statement says simply that λ is the order of the subgroup leaving P_1 fixed, which is immediate from the theorem.

COROLLARY 2. (Kummer's definition of ideal prime factors of cyclotomic integers [**Ku**, p. 322]) *Let p, n be positive integers, with p prime, $n > 2$, $p \nmid n$. Let K be the subfield of Λ_n invariant under the automorphism $\alpha \mapsto \alpha^p$. There is an element $\psi \in K$ whose norm, as an element of $K \supset \mathbf{Q}$, is divisible by p but not by p^2. For any such ψ, let ψ_1, ψ_2, $\ldots$, ψ_e, where $e = [K : \mathbf{Q}]$, be its conjugates in K, and let $\Psi_i = \prod_{j \neq i} \psi_j$.*

Then $[p] = \prod_{i=1}^{e}[p, \psi_i]$, the divisors $[p, \psi_i]$ are prime in Λ_n, and $[p, \psi_i]^\mu$ divides $f(\alpha) \in \Lambda_n$ if and only if $p^\mu | f(\alpha)\Psi_i^\mu$.

DEDUCTION: By the choice of K, $\lambda = 1$ in Corollary 1. Therefore, $[p]$ is a product of $[K:\mathbf{Q}] = e$ distinct divisors prime in K, say $[p] = P_1 P_2 \cdots P_e$, and the P_i are conjugate under the Galois group of $K \supset \mathbf{Q}$. By Theorem 2, $P_1 = [p, \psi]$ for some $\psi \in K$. When Corollary 1 is applied with Λ_n in place of K, λ is the least positive integer, Kummer denotes it by f, for which $p^f \equiv 1 \bmod n$. In other words, f is the order of the subgroup of the Galois group of $\Lambda_n \supset K$ generated by $\alpha \mapsto \alpha^p$. This subgroup corresponds to the subfield K of Corollary 2, so $f = [\Lambda_n : K]$ by basic Galois theory. It follows from Corollary 1 that $[p]$ is a product of $[\Lambda:\mathbf{Q}]/[\Lambda_n:K] = [K:\mathbf{Q}] = e$ divisors prime in Λ_n. Therefore, the factors of $[p] = \prod_i[p, \psi_i]$ are prime in Λ_n and the remaining statements of the corollary follow easily.

§2.12 Quadratic Reciprocity.

Let q be a prime integer greater than 2, and let Λ_q be the splitting field of $x^q - 1$ over $\mathbf{Q}$. Since the Galois group G of Λ_q over $\mathbf{Q}$ is cyclic of order $q - 1$ (that is, there is† a primitive root mod q), Λ_q contains a unique quadratic extension K of $\mathbf{Q}$, namely, the subfield of Λ_q corresponding to the subgroup of *squares* in G. By the Corollary of §2.11, the factorization of $[p]$ in K for p a prime, $p \neq q$, contains two factors if p is a square mod q (so $\lambda = 1$); otherwise, $[p]$ is prime in K.

---

†By Fermat's theorem, $x^q - x$ has q distinct roots in the field $\mathbf{F}_q$. For any proper divisor d of $q - 1$, $x^{q-1} - 1 = \Phi_{q-1}(x)(x^d - 1)Q(x)$, where Q is a polynomial with coefficients in $\mathbf{Z}$ and $\Phi_{q-1} = N_{\Lambda_{q-1}}(x - \alpha)$ is the polynomial with coefficients in $\mathbf{Z}$ of which primitive $(q-1)$st roots of unity are roots. ($x^d - 1$ divides $x^{q-1} - 1$ and is relatively prime to $\Phi_{q-1}(x)$.) Since $x^q - x = x\Phi_{q-1}(x)(x^d - 1)Q(x)$ has q distinct roots in $\mathbf{F}_q$, $\Phi_{q-1}(x)$ and $x^d - 1$ have no roots in common. Since d was arbitrary, the roots of $\Phi_{q-1}(x)$ in $\mathbf{F}_q$ are primitive $(q-1)$st roots of unity. Therefore, there are elements of the multiplicative group of $\mathbf{F}_q$ of order $q - 1$.

On the other hand, K has the form $K = \mathbf{Q}(\sqrt{d})$ for some squarefree $d \in \mathbf{Z}$, and the factorization of $[p]$ in K can be determined by the method of §2.5. Specifically†, $K = \mathbf{Q}(\sqrt{q})$ when $q \equiv 1 \bmod 4$ and $K = \mathbf{Q}(\sqrt{-q})$ when $q \equiv 3 \bmod 4$. Let q^* denote q when $q \equiv 1 \bmod 4$ and $-q$ when $q \equiv 3 \bmod 4$. Then, for $p \neq 2$, $[p]$ is a product of two divisors in $K = \mathbf{Q}(\sqrt{q^*})$ when q^* is a square mod p, and is prime in K otherwise. Therefore:

If p and q are primes distinct from one another and distinct from 2, then p is a square mod q if and only if q^* is a square mod p, where q^* is defined as above.

---

†Gauss, D. A., §356. Let $\theta_0 = \sum \alpha^j$, where j runs over all nonzero squares mod q and let $\theta_1 = \sum \alpha^j$, where j runs over nonsquares. Then $1 + \theta_0 + \theta_1 = 1 + \alpha + \alpha^2 + \cdots + \alpha^{q-1}$, which is 0 because $x - \alpha$ divides $x^q - 1$ but not $x - 1$, so it divides $(x^q - 1)/(x - 1) = x^{q-1} + x^{q-2} + \cdots + 1$. Both θ_0 and θ_1 are elements of K integral over $\mathbf{Z}$, so $\theta_0\theta_1$ is integral over $\mathbf{Z}$. $\theta_0\theta_1$ is invariant under the Galois group and is therefore in $\mathbf{Q}$. Thus, $\theta_0\theta_1 \in \mathbf{Z}$. Both θ_0 and θ_1 are sums of $(q-1)/2$ terms, so, before terms are combined, $\theta_0\theta_1$ is a sum of $(q-1)^2/4$ terms, each a power of α. If α^{-1} is a term of θ_0—that is, if -1 is a square mod q—then 1 does not occur among the $(q-1)^2/4$ terms of $\theta_0\theta_1$, and, by symmetry, each of the $(q-1)$ other powers of α occurs $(q-1)/4$ times. Therefore, $\theta_0\theta_1 = (\theta_0 + \theta_1)(q-1)/4 = (1-q)/4$. In particular, in this case $q \equiv 1 \bmod 4$. If α^{-1} is a term of θ_1, then 1 occurs $(q-1)/2$ times in $\theta_0\theta_1$, and the remaining $((q-1)^2/4) - (q-1)/2 = (q^2 - 4q + 3)/4 = (q-1)(q-3)/4$ terms of $\theta_0\theta_1$ are evenly distributed among the other $q - 1$ powers of α. Thus, $\theta_0\theta_1 = ((q-1)/2) + (\theta_0 + \theta_1)(q-3)/4 = (2q - 2 - q + 3)/4 = (q+1)/4$. In particular, in this case $q \equiv -1 \bmod 4$. Thus, $(x - \theta_0)(x - \theta_1) = x^2 + x + \theta_0\theta_1 = ((2x + 1)^2 + 4\theta_0\theta_1 - 1)/4 = ((2x + 1)^2 - q^*)/4$, where $q^* = q$ if $q \equiv 1 \bmod 4$ and $q^* = -q$ if $q \equiv 3 \bmod 4$. Since $\mathbf{Q}(\sqrt{q^*})$ and K are both splitting fields of $(x - \theta_0)(x - \theta_1)$, they are isomorphic.

Part 3: Applications to the Theory of Algebraic Curves

§3.1 Function Fields.

Let $\mathbf{Q}[x]$ denote the ring of polynomials in an indeterminate x with coefficients in the field $\mathbf{Q}$ of rational numbers. An *algebraic function field in one variable over* $\mathbf{Q}$ is an extension of $\mathbf{Q}[x]$ of finite degree. For short, such fields will be called *function fields.* Note that function fields are to $\mathbf{Q}[x]$ what algebraic number fields are to $\mathbf{Z}$. Since $\mathbf{Q}[x]$ is a natural ring (§1.2), divisor theory applies* to function fields.

The field of quotients $\mathbf{Q}(x)$ of $\mathbf{Q}[x]$ is called "the field of rational functions in x with coefficients in $\mathbf{Q}$." In the 19th century, it was usual to call the elements of a function field "functions", but in the 20th century the word "function" has become so firmly fixed in its set-theoretic meaning that to use it in any other way risks confusion and misunderstanding. For this reason, the word function will be used here only in the phrase "function field" and, occasionally, the phrase† "rational function".

§3.2 Parameters and Constants.

Let u be an element of a function field $K \supset \mathbf{Q}(x)$. Since u is algebraic over $\mathbf{Q}(x)$, $a_0(x)u^n + a_1(x)u^{n-1} + \cdots + a_n(x) = 0$, where the $a_i(x)$ are in $\mathbf{Q}[x]$ and $a_0(x) \neq 0$. Moreover, as was

---

*Function fields are also extensions of finite degree of the natural ring $\mathbf{Z}[x]$; consequently, divisor theory applies in *another* way to them. These two divisor theories are different—for example, integers greater than 1 are units in $\mathbf{Q}[x]$ but not in $\mathbf{Z}[x]$. Only the theory stemming from $\mathbf{Q}[x]$ will be considered.

†A "rational function" in indeterminates $x_1, x_2, \ldots, x_n$ is an element of the quotient field of the natural ring $\mathbf{Q}[x_1, x_2, \ldots, x_n]$. In the case of *one* indeterminate, such a quotient of polynomials can be regarded as a function of a complex variable in the set-theoretic sense, provided ∞ is admitted as an element of the range. This interpretation depends on the fact that no complex number is simultaneously a root of two relatively prime polynomials in one indeterminate.

seen in §1.9, one can assume without loss of generality that
$h(X) = a_0(x)X^n + a_1(x)X^{n-1} + \cdots + a_n(x)$ is irreducible as a
polynomial in X with coefficients in $\mathbf{Q}[x]$, in which case $h(X)$
is determined up to multiplication by a unit of $\mathbf{Q}[x]$, that is,
up to multiplication by a nonzero rational number. If the $a_i(x)$
are all independent of x, u is called a *constant* of K. Otherwise,
u is a *parameter* of K.

If u is a parameter of K, the homomorphism s from the
natural ring $\mathbf{Q}[X]$ to K defined by $s(X) = u$ and $s(c) = c$ for
$c \in \mathbf{Q}$ is one-to-one. (If $g(X) \in \mathbf{Q}[X]$ satisfied $g(u) = 0$, then
$h(X)$ would divide $g(X)$ as a polynomial with coefficients in
$\mathbf{Q}[x] \supset \mathbf{Q}$. Since g can be assumed to be monic, so can h and
g/h. Then the theorem of Part 0 implies that the coefficients
of h are roots of polynomials with coefficients in $\mathbf{Q}$, which implies that the coefficients of h have degree 0 in x, contrary to
assumption.) Thus, the image of s is a subring of K isomorphic to the natural ring $\mathbf{Q}[X]$. Let $\mathbf{Q}[u]$ denote the image of
s. The field of quotients $\mathbf{Q}(u)$ of $\mathbf{Q}[u]$ is a subfield of K isomorphic to the field of rational functions in one indeterminate.
The field extension $K \supset \mathbf{Q}(u)$ is of finite degree, because the
relation $a_0(x)u^n + a_1(x)u^{n-1} + \cdots + a_n(x) = 0$ can be rewrit-
ten in the form $b_0(u)x^m + b_1(u)x^{m-1} + \cdots + b_m(u) = 0$, where
$m > 0$ and $b_0 \neq 0$ (when u is not a constant of K), which
shows that x is algebraic over $\mathbf{Q}(u)$; thus, $[\mathbf{Q}(x, u): \mathbf{Q}(u)] <$
∞, which gives $[K : \mathbf{Q}(u)] = [K : \mathbf{Q}(x, u)][\mathbf{Q}(x, u): \mathbf{Q}(u)] =$
$[K : \mathbf{Q}(x)][\mathbf{Q}(x, u): \mathbf{Q}(x)]^{-1}[\mathbf{Q}(x, u): \mathbf{Q}(u)] \leq [K : \mathbf{Q}(x)] \cdot 1 \cdot$
$[\mathbf{Q}(x, u): \mathbf{Q}(u)] < \infty$.

In short, if u is any parameter of K, then $\mathbf{Q}(u)$ is a subfield of K isomorphic to the field of rational functions in one
indeterminate, and $K \supset \mathbf{Q}(u)$ is an extension of finite degree.

§3.3.

Let K be a function field, and let u be a parameter of K. As
an extension of finite degree of the natural ring $\mathbf{Q}[u]$, K has a
divisor theory. However, this divisor theory in K depends on

the choice of the parameter u. It will be shown in the next article that there is a natural way to define a *global* divisor theory—one not dependent on the choice of the parameter—using the following lemma.

Notation: Let K be a function field, u a parameter in K, and f and g polynomials with coefficients in K. Let $[f]_u|[g]_u$ mean that, in the divisor theory of K as an extension of the natural ring $\mathbf{Q}[u]$, the divisor represented by f divides the divisor represented by g.

LEMMA. *If x and u are parameters of a function field K, and if x is integral over $\mathbf{Q}[u]$, then $[f]_x|[g]_x$ implies $[f]_u|[g]_u$.*

PROOF: $[f]_x|[g]_x$ means $fq = g\pi$, where q has coefficients in K integral over $\mathbf{Q}[x]$ and π has coefficients $v_1, v_2, \ldots, v_n$ in $\mathbf{Q}[x]$ of which 1 is a g. c. d. By the Euclidean algorithm in $\mathbf{Q}[x]$, there exist $a_1, a_2, \ldots, a_n$ in $\mathbf{Q}[x]$ such that $a_1 v_1 + a_2 v_2 + \cdots + a_n v_n = 1$. In the divisor theory of the extension $K \supset \mathbf{Q}[u]$, $[\pi]_u|[1]_u$ because $[\pi]_u|[v_i]_u$ ((9) of §1.11) and $[v_i]_u|[a_i v_i]_u$ (from the definition, because a_i is a sum of elements of K integral over $\mathbf{Q}[u]$) which implies $[\pi]_u|[a_i v_i]_u$ ((3) of §1.8), which in turn implies $[\pi]_u|[a_1 v_1 + a_2 v_2 + \cdots + a_n v_n]_u = [1]_u$ ((1) of §1.8). Since the coefficients v_i of π are in $\mathbf{Q}[x]$, they are integral over $\mathbf{Q}[x]$, which implies‡ that they are integral over $\mathbf{Q}[u]$, that is, $[1]_u|[\pi]_u$. Thus, $[\pi]_u = [1]_u$, and $[f]_u^{-1}[g]_u = [f]_u^{-1}[g\pi]_u = [f]_u^{-1}[fq]_u = [q]_u$. Since the coefficients of q are integral over $\mathbf{Q}[x]$, they are integral over $\mathbf{Q}[u]$, $[q]_u$ is integral, and $[f]_u|[g]_u$, as was to be shown.

§3.4 Global Divisors.

Let K be a function field, let u and x be parameters of K, and let $\mathcal{D}_u$ and $\mathcal{D}_x$ denote the groups of nonzero divisors in K as an extension of the natural rings $\mathbf{Q}[u]$ and $\mathbf{Q}[x]$ respectively. The lemma of §3.3 implies that, if x is integral over $\mathbf{Q}[u]$,

---

‡See Proposition 2 of Part 0.

there is a natural onto homomorphism $\mathcal{D}_x \to \mathcal{D}_u$, namely, the homomorphism which sends $[f]_x \in \mathcal{D}_x$ to $[f]_u \in \mathcal{D}_u$. This homomorphism will be called *restriction from x to u*.

DEFINITION. A *global divisor* in a function field K is an assignment A, to each parameter u of K, of a nonzero divisor A_u in $\mathcal{D}_u$, called the *restriction of A to u*, in such a way that $A_u \in \mathcal{D}_u$ is the restriction from x to u of $A_x \in \mathcal{D}_x$ whenever this restriction is defined, that is, whenever x is integral over $\mathbf{Q}[u]$.

(Note that, according to this definition, zero is not a global divisor. One might also define a zero global divisor, but then the global divisors would not be a group.)

If f is a nonzero polynomial with coefficients in K, then $[f]_u$ represents an element of $\mathcal{D}_u$ for each parameter u of K. By the lemma of §3.3, $u \mapsto [f]_u$ is a global divisor in K. This global divisor will be denoted $[f]$.

It was shown in §1.23 that any A_u in $\mathcal{D}_u$ can be written in one and only one way as a quotient $A_u = B_u/C_u$, in which B_u and C_u are relatively prime integral divisors in $\mathcal{D}_u$. Let A be a global divisor in K, and let B and C assign to each parameter u of K the relatively prime integral divisors B_u and C_u, respectively, such that $A_u = B_u/C_u$. Then B and C are global divisors, as can be seen as follows. Let x and u be parameters of K, with x integral over $\mathbf{Q}[u]$, and let $\iota(B_x)$ and $\iota(C_x)$ denote the restrictions of these divisors from x to u. Then $\iota(B_x)$ and $\iota(C_x)$ are integral (by the lemma, $[1]_x|B_x$ implies $[1]_u|\iota(B_x)$ and similarly $[1]_u|\iota(C_x)$) and relatively prime ($[B_x, C_x]_x|[1]_x$ implies $[\iota(B_x), \iota(C_x)]_u|[1]_u$) and have quotient A_u ($A_x C_x = B_x$ implies $A_u \iota(C_x) = \iota(B_x)$ because A is a global divisor and ι is a homomorphism). In short, $B_u = \iota(B_x)$ and $C_u = \iota(C_x)$, as was to be shown. The global divisors B and C defined in this way will be called the *numerator* and *denominator* of A, respectively.

DEFINITION. A global divisor is *integral* if its denominator is [1], or, what is the same, if its restriction to u is integral for all parameters u of K.

Global divisors in K obviously form a commutative *group*. Let $\mathcal{D}_K$ denote this group. The integral global divisors are a subset of $\mathcal{D}_K$ closed under multiplication, and every element of $\mathcal{D}_K$ can be written as a quotient of integral elements. One global divisor will be said to *divide* another if their quotient is integral. Clearly, two global divisors are equal if and only if each divides the other.

If A, B, $\ldots$, E are global divisors in K, let $[A, B, \ldots, E]$ assign to each parameter u of K the divisor $[A_u, B_u, \ldots, E_u]_u$ in $\mathcal{D}_u$. Then $[A, B, \ldots, E]$ is a global divisor in K (the restriction of a g.c.d. is the g.c.d. of the restrictions) which divides A, B, $\ldots$, E and is divisible by any global divisor which divides A, B, $\ldots$, E. Thus, $[A, B, \ldots, E]$ is the greatest common divisor of A, B, $\ldots$, E. In particular, $[f]$ is the greatest common divisor of the global divisors $[\alpha]$, where α runs over the coefficients of f. (However, not every global divisor can be expressed in the form $[f]$ where f is a polynomial with coefficients in K—see §3.23.)

§3.5.

In the remainder of this book, "divisor" will mean *global* divisor in a function field.

PROPOSITION. *Let A be a divisor in a function field K and let x be a parameter of K. If the restrictions A_x and $A_{1/x}$ of A to x and $1/x$ are both integral, then A is integral.*

COROLLARY. *If A and B are divisors in K, if x is a parameter of K, if $A_x = B_x$, and if $A_{1/x} = B_{1/x}$, then $A = B$.*

DEDUCTION: Since $A_x | B_x$ and $A_{1/x} | B_{1/x}$, both $A_x^{-1} B_x$ and $A_{1/x}^{-1} B_{1/x}$ are integral, so $A^{-1} B$ is integral by the proposition, that is, A divides B. In the same way, B divides A, so $A = B$.

LEMMA. *Let x and y be parameters in a function field K. There is a nonzero polynomial $F(X,Y) \in \mathbf{Q}[X,Y]$ in two indeterminates such that $F(x,y) = 0$ in K. If $F(X,Y) = a_0(Y)X^n + a_1(Y)X^{n-1} + \cdots + a_n(Y)$ is such a polynomial, and if A is a divisor in K for which A_x is integral, then the denominator of A_y divides $[a_0(y)^j]_y$ for all sufficiently large integers j.*

PROOF: The existence of $F(X,Y)$ was shown in §3.2 (y is algebraic over $\mathbf{Q}[x]$).

Let $z \in K$ be defined by $z = y + (1/a_0(y))$, that is, $ya_0(y) - za_0(y) + 1 = 0$. Then y is integral over $\mathbf{Q}[z]$, because the coefficient of the highest power of y in $ya_0(y) - za_0(y) + 1$ is independent of z (it is the coefficient of the leading term of $a_0(y)$). Since elements of K integral over $\mathbf{Q}[z]$ form a ring, both $a_0(y)$ and $a_0(y)^{-1} = z - y$ are integral over $\mathbf{Q}[z]$. Thus, both $[a_0(y)]_z$ and $[1/(a_0(y))]_z$ are integral (in $\mathcal{D}_z$), which is to say that $[a_0(y)]_z = [1]_z$. On the other hand, multiplication of $a_0(y)x^n + \cdots = 0$ by $a_0(y)^{n-1}$ shows that $a_0(y)x$ is integral over $\mathbf{Q}[y]$, that is, $[1]_y|[a_0(y)x]_y$. Since y is integral over $\mathbf{Q}[z]$, the lemma of §3.3 implies that $[1]_z|[a_0(y)x]_z = [a_0(y)]_z[x]_z = [x]_z$, that is, x is integral over $\mathbf{Q}[z]$. Therefore, A_z is the restriction to z of A_x. Since A_x is integral by assumption, A_z is integral.

Let f be a polynomial with coefficients in K such that $A_y = [f]_y$. Since y is integral over $\mathbf{Q}[z]$, $[f]_z$ is A_z, which was just shown to be integral. Therefore, every coefficient α of f is integral over $\mathbf{Q}[z]$, which is to say that α satisfies a relation of the form $\alpha^\mu = \sum_{i=1}^\mu c_i(z)\alpha^{\mu-i}$, where $c_i(z) \in \mathbf{Q}[z]$. Because $a_0(y)z = a_0(y)y + 1$ can be expressed as a polynomial in y, $a_0(y)^k c_i(z)$ can be expressed as a polynomial in y whenever $k \geq \deg c_i$. The relation $(a_0(y)^j \alpha)^\mu = \sum_{i=1}^\mu c_i(z)a_0(y)^{ij}(a_0(y)^j\alpha)^{\mu-i}$ then shows that $a_0(y)^j\alpha$ is integral over $\mathbf{Q}[y]$ for all sufficiently large j. (Specifically, if $ij \geq \deg c_i$ for $i = 1, 2, \ldots, \mu$, then $a_0(y)^j\alpha$ is integral over $\mathbf{Q}[y]$.) Since this is true for each coefficient α of f,

$[a_0(y)^j f]_y$ is integral for all sufficiently large j. But $[f]_y = A_y$. Therefore, $[a_0(y)]_y^j A_y$ is integral for all sufficiently large j, that is, the denominator of A_y divides $[a_0(y)]_y^j$ for large j, as was to be shown.

PROOF OF THE PROPOSITION: Let A be a divisor in K such that A_x and $A_{1/x}$ are integral. It is to be shown that A_y is integral for all parameters y of K.

Consider first the special case in which $y = (x - c)^{-1}$, where $c \in \mathbf{Q}$. Then $yx - cy - 1 = 0$ and, by the lemma, the denominator of A_y divides $[y^j]_y$ for all sufficiently large j. On the other hand, $(cy + 1)x^{-1} - y = 0$, and, by the lemma, the denominator of A_y divides $[(cy + 1)^j]_y$ for all sufficiently large j (because $A_{1/x}$ is integral). But, for any c, $[y, cy + 1]_y = [1]_y$, which implies $[y^j, (cy + 1)^j]_y = [1]_y$ for all $j > 0$. Therefore, the denominator of A_y divides $[1]_y$, that is, A_y is integral, as was to be shown.

Now let y be an arbitrary parameter of K. Let $a_0(y)x^n + a_1(y)x^{n-1} + \cdots + a_n(y) = 0$, where $a_i(y) \in \mathbf{Q}[y]$ and $a_0(y) \neq 0$. Since this relation can be divided by any common factor of the $a_i(y)$, one can assume without loss of generality that $[a_0(y), a_1(y), \ldots, a_n(y)]_y = [1]_y$. Let $v = (x - c)^{-1}$, where $c \in \mathbf{Q}$. As was just shown, A_v is integral. A relation satisfied by v and y can be found by setting $x = c + v^{-1}$ in $\sum a_i(y)x^{n-i} = 0$ and multiplying by v^n, which gives $(a_0(y)c^n + a_1(y)c^{n-1} + \cdots + a_n(y))v^n + (\text{terms of lower degree in } v) = 0$. By the lemma, the denominator of A_y divides $[a_0(y)c^n + a_1(y)c^{n-1} + \cdots + a_n(y)]_y^j$ for any $c \in \mathbf{Q}$, provided j is sufficiently large. Let $b_i(y) = a_0(y)i^n + a_1(y)i^{n-1} + \cdots + a_n(y)$ for $i = 1, 2, \ldots, n+1$. Since the $(n + 1) \times (n + 1)$ matrix of coefficients (i^μ) $(1 \leq i \leq n + 1, 0 \leq \mu \leq n)$ has nonzero determinant (a Vandermonde determinant), the $a_i(y)$ can be written as linear combinations of the $b_i(y)$ with rational coefficients, and $[b_0(y), b_1(y), \ldots, b_n(y)]_y = [a_0(y), a_1(y), \ldots, a_n(y)]_y = [1]_y$. Therefore, $[b_0(y)^j, b_1(y)^j, \ldots, b_n(y)^j]_y = [1]_y$ for all j and the denominator of A_y

divides $[1]_y$, as was to be shown.

§**3.6.**

DEFINITIONS. Let B be an integral divisor in a function field K. A divisor A *avoids* B if both numerator and denominator of A are relatively prime to B. A divisor A is *concentrated at* B if both numerator and denominator of A divide some power of B.

PROPOSITION. *Let B be an integral divisor in a function field K. Any divisor in K can be written, in just one way, as a product of a divisor which avoids B and a divisor concentrated at B. In other words, the group of divisors is the direct product of the subgroup of divisors which avoid B and the subgroup of divisors concentrated at B.*

PROOF: Consider first the case in which the given divisor A is integral. Let $C^{(j)} = A/[A, B^j]$. Then $C^{(j)}$ is an integral divisor, $[A, B^j]$ is an integral divisor concentrated at B, and A is their product. It will be shown that $C^{(j)}$ is relatively prime to B for j sufficiently large, that is, $[C^{(j)}, B] = [1]$ for j sufficiently large. Let x be a parameter of K. By Theorem 1, A_x and B_x can be written as products of (positive) powers of relatively prime integral divisors $E_x^{(i)}$ in $\mathcal{D}_x$, say $A_x = \prod (E_x^{(i)})^{\sigma_i}$ and $B_x = \prod (E_x^{(i)})^{\tau_i}$. The exponent of $E_x^{(i)}$ in $C_x^{(j)}$ is then $\sigma_i - \min(\sigma_i, j\tau_i)$. Let j_1 be large enough that $j_1 \tau_i \geq \sigma_i$ whenever $\tau_i > 0$. Then $j > j_1$ implies that the exponent of $E_x^{(i)}$ in $C_x^{(j)}$ is zero whenever $\tau_i > 0$. Thus, $C_x^{(j)}$ is a product of divisors $E_x^{(i)}$ relatively prime to B_x, which implies $[C_x^{(j)}, B_x]_x = [1]_x$. In the same way, there is an integer j_2 such that $j > j_2$ implies $[C_{1/x}^{(j)}, B_{1/x}]_{1/x} = [1]_{1/x}$. By the proposition of §3.5, $j > \max(j_1, j_2)$ implies $[C^{(j)}, B] = [1]$. Therefore, $A = C^{(j)}[A, B^j]$ is a factorization of the desired type when $j > \max(j_1, j_2)$. This factorization is unique, because if $A = A^{(1)} A^{(2)}$, where $A^{(1)}$ is relatively prime to B and

$A^{(2)}|B^k$ for some k, then $[A, B^j] = [A^{(2)}, B^j] = A^{(2)}$ whenever $j \geq k$, which shows that the factorization $A = A^{(1)}A^{(2)}$ coincides with the factorization found above.

To write an arbitrary divisor as a product of a divisor which avoids B and a divisor which is concentrated at B, apply the above construction to its numerator and denominator.

§3.7.

PROPOSITION. *The restriction map from $\mathcal{D}_K$ to $\mathcal{D}_x$ has as its kernel the divisors concentrated at the denominator of $[x]$.*

PROOF: Let $E \in \mathcal{D}_K$ be the denominator of $[x]$. Let $A = B/C$ be a divisor in K, where B and C are relatively prime integral divisors. If A is concentrated at E, then B and C divide E^j for large j. Therefore, B_x and C_x divide E_x^j for large j. Since $[x]_x$ is integral, $E_x = [1]_x$. Therefore, B_x and C_x, being integral divisors which divide $[1]_x^j$, must both be $[1]_x$, that is, A must be in the kernel of $\mathcal{D}_K \to \mathcal{D}_x$.

Conversely, if $A_x = [1]_x$, then $B_x/C_x = [1]_x$, where B_x and C_x are integral and relatively prime. Therefore, $B_x = C_x = [1]_x$. Let $y = 1/x$, that is, $yx - 1 = 0$. Since B^{-1} and C^{-1} have integral restrictions to x, the denominators of B_y^{-1} and C_y^{-1} divide $[y]_y^j$ for j sufficiently large (Lemma, §3.5). But B_y and C_y are the denominators of B_y^{-1} and C_y^{-1}, respectively, and $E_y = [y]_y$. Thus, B_y and C_y divide E_y^j for large j. Since B_x and C_x also divide E_x^j for large j (all three are $[1]_x$), B and C are concentrated at E by the proposition of §3.5, as was to be shown.

DEFINITION. *A parameter x in a function field K supports a divisor A in K if A avoids the denominator of $[x]$.*

COROLLARY. *The restriction map $\mathcal{D}_K \to \mathcal{D}_x$ is an isomorphism from the subgroup of $\mathcal{D}_K$ of divisors supported by x to $\mathcal{D}_x$. In other words, any $A_x \in \mathcal{D}_x$ is the restriction to x of some* divisor $A \in \mathcal{D}_K$ supported by x, and two divisors supported by x which have the same restriction to x are equal.

DEDUCTION: As was shown in §3.6, $\mathcal{D}_K$ is the direct product of divisors concentrated at the denominator of $[x]$ and divisors which avoid the denominator of $[x]$. The former are, by the proposition, the kernel of $\mathcal{D}_K \to \mathcal{D}_x$, and the latter are by definition the divisors supported by x. Every element of $\mathcal{D}_x$ can be written in the form $[f]_x$ for some polynomial f with coefficients in K. Since $[f]_x$ is the image of $[f]$ under $\mathcal{D}_K \to \mathcal{D}_x$, this homomorphism is onto, and the corollary follows.

§3.8.

PROPOSITION. *Given a divisor A in a function field K, and given a parameter x of K, A is supported by $u = (x-a)^{-1}$ for all but a finite number of rational numbers a.*

PROOF: Since a quotient of divisors supported by u is supported by u, it will suffice to prove the proposition in the case in which A is integral. By Lemma (8) of §1.10, $A_x||[\phi(x)]_x$ for some $\phi(x) \in \mathbf{Q}[x]$. If $a \in \mathbf{Q}$ is not a root of ϕ, then $x - a$ does not divide ϕ. Since $\mathbf{Q}[x]$ is a natural ring and $x - a$ is irreducible (prime) in $\mathbf{Q}[x]$, it follows that $[x-a]_x$ and $[\phi(x)]_x$ are relatively prime. Since $\mathbf{Q}[x] = \mathbf{Q}[x-a]$, restriction to x is the same as restriction to $x - a$, and $[x-a]_{x-a}$ and A_{x-a} are relatively prime. Thus, $[1/u]_{1/u}$ and $A_{1/u}$ are relatively prime, that is, the restrictions to $1/u$ of A and the denominator of $[u]$ are relatively prime. Since the same is trivially true of their restrictions to u, A is supported by $u = (x-a)^{-1}$ whenever $\phi(a) \neq 0$.

An integral divisor in K is *reducible* in K if it can be written as a product of two integral divisors, neither of which is $[1]$; otherwise it is *irreducible* in K. An integral divisor A in K is *prime* in K if the integral divisors in K not divisible by A form a nonempty set closed under multiplication. (See §1.18.)

COROLLARY. *An integral divisor other than* [1] *is irreducible in K if and only if it is prime in K.*

DEDUCTION: If A is prime, then $A \neq [1]$. Moreover, if $A = BC$, then A divides B or C, say $B = AD$; then $[1] = DC$, $C = [1]$, which shows that A is irreducible. Suppose now that $A \neq [1]$ is integral, that A is irreducible, and that $A|BC$, where B and C are integral. By the proposition, there is a parameter u which supports A, B, and C. Then A_u is irreducible (if $A_u = D_u E_u$, then D_u or E_u are restrictions to u of divisors D and E supported by u, and $A = DE$ by the corollary of §3.7, so either D or E is [1] and either D_u or E_u is $[1]_u$). Therefore A_u is prime (§1.18), so $A_u|B_u$ or $A_u|C_u$; by the corollary of §3.7 it follows that $A|B$ or $A|C$, so A is prime.

§3.9.

Notation: For a nonzero polynomial f with coefficients in K, let $\{f\}$ denote the numerator of $[f]$. If $\alpha_1, \alpha_2, \ldots, \alpha_\nu \in K$ are not all zero, let $\{\alpha_1, \alpha_2, \ldots, \alpha_\nu\}$ denote $\{f\}$, where f is a polynomial whose coefficients are $\alpha_1, \alpha_2, \ldots, \alpha_\nu$. (Since $[f]$ is independent of the choice of f, so is $\{f\}$.) Note that $\{1/x\}$ is the denominator of $[x]$. Thus, x supports A if and only if A avoids $\{1/x\}$.

PROPOSITION. *Among integral divisors, $\{\alpha_1, \alpha_2, \ldots, \alpha_\nu\}$ is the greatest common divisor of $\{\alpha_1\}, \{\alpha_2\}, \ldots, \{\alpha_\nu\}$; that is, $\{\alpha_1, \alpha_2, \ldots, \alpha_\nu\}|\{\alpha_i\}$ for each i, and any integral divisor which divides $\{\alpha_i\}$ for each i divides $\{\alpha_1, \alpha_2, \ldots, \alpha_\nu\}$.*

PROOF: Let A be an integral divisor which divides $\{\alpha_i\}$ for $i = 1, 2, \ldots, \nu$. It is to be shown that, for each parameter x of K, A_x divides the numerator of $[\alpha_1, \alpha_2, \ldots, \alpha_\nu]_x$, which in turn divides the numerator of $[\alpha_i]_x$ for all i. By Theorem 1, there are relatively prime integral divisors $B_1, B_2, \ldots, B_\mu$ in $\mathcal{D}_x$ such that A_x and the $[\alpha_i]_x$ are all products of powers of the B_i, say $A_x = B_1^{a_1} B_2^{a_2} \cdots B_\mu^{a_\mu}$ and $[\alpha_i]_x = B_1^{b_{i1}} B_2^{b_{i2}} \cdots B_\mu^{b_{i\mu}}$. Let f be a polynomial whose coefficients are $\alpha_1, \alpha_2, \ldots, \alpha_\nu$. Then $[f]_x = B_1^{c_1} B_2^{c_2} \cdots B_\mu^{c_\mu}$, where $c_j = \min(b_{1j}, b_{2j}, \ldots, b_{\nu j})$ (because this is the g. c. d. of the $[\alpha_i]_x$). Thus, the numerator of $[f]_x$ is $B_1^{d_1} B_2^{d_2} \cdots B_\mu^{d_\mu}$, where $d_j = \max(c_j, 0)$, and the numerator of $[\alpha_i]_x$ is $B_1^{e_{i1}} B_2^{e_{i2}} \cdots B_\mu^{e_{i\mu}}$, where $e_{ij} = \max(b_{ij}, 0)$. Because $A|\{\alpha_i\}$, $a_j \leq e_{ij}$ for all i. For fixed j, if some $b_{ij} < 0$, $a_j \leq e_{ij} = 0 \leq d_j$ follows; if all $b_{ij} \geq 0$, then $a_j \leq e_{ij} = b_{ij}$, and $a_j \leq c_j = d_j$ follows. Thus, A_x divides the numerator of $[f]_x$. On the other hand, since $c_j \leq b_{ij}$ for all i, $d_j \leq e_{ij}$ for all i, that is, the numerator of $[f]_x$ divides the numerator of $[\alpha_i]_x$ for all i.

§3.10.

PROPOSITION (Analog of Theorem 2). *Given two integral divisors A and B in a function field K, there is a parameter x of K, supporting both A and B, such that $A|\{x\}$ and $A^{-1}\{x\}$ is relatively prime to B.*

PROOF: Let u be a parameter of K which supports AB. By Theorem 2, there is an $x \in K$ integral over $\mathbf{Q}[u]$ such that A_u divides $[x]_u$ and $A_u^{-1}[x]_u$ is relatively prime to B_u. Since the denominator of $[x]_u$ is $[1]_u$, the denominator of $[x]$ is concentrated at the denominator of $[u]$ (§3.7), that is, $\{1/x\}$ divides a power of $\{1/u\}$. Since AB avoids $\{1/u\}$, it therefore avoids $\{1/x\}$, which implies that x supports both A and B. The denominator of $A^{-1}\{x\}$ has restriction $[1]_u$ to u ($A_u|[x]_u$) and is supported by u (it divides A), so the denominator of $A^{-1}\{x\}$ is $[1]$ (§3.7, Corollary), that is, $A|\{x\}$. In the same way, $[A^{-1}\{x\}, B]$ has restriction $[1]_u$ to u and is supported by u, which implies $[A^{-1}\{x\}, B] = [1]$.

COROLLARIES. (1) *If A is an integral divisor in a function field K, and if x is a parameter of K for which $A|\{x\}$, then $A = \{x, y\}$ for some $y \in K$. (2) Any divisor in a function field K can be written in the form $\{x, y\}/\{u, v\}$, where x, y, u, v are nonzero elements of K.*

DEDUCTIONS: (1) By the proposition, there is a parameter y of K with $A|\{y\}$ and with $A^{-1}\{y\}$ relatively prime to $A^{-1}\{x\}$. Thus, $\{x, y\} = AC$ for some integral divisor C (A is an integral divisor which divides both $\{x\}$ and $\{y\}$). Since C divides both $A^{-1}\{x\}$ and $A^{-1}\{y\}$, $C = [1]$. (2) Any divisor can be written B/C, where B and C are integral divisors. By the proposition, there are parameters x and u of K such that $B|\{x\}$ and $C|\{u\}$. By the first corollary, $B = \{x, y\}$ and $C = \{u, v\}$, where y, $v \in K$ are parameters of K and are therefore nonzero.

Corollary (2) shows that, as in the basic theory of divisors, the theory of (global) divisors in function fields is *independent of the ambient field* in the sense that $K \subset L$ implies $\mathcal{D}_K \subset \mathcal{D}_L$ in a natural way. For, if x, y, u, v are nonzero elements of K, they are also nonzero elements of L, and $\{x, y\}/\{u, v\}$ represents a divisor in L as well as a divisor in K. Since operations in $\mathcal{D}_K$ (multiplication, inversion, taking numerators, finding greatest common divisors, etc.) depend only of the field operations in K, they are unaffected if K is replaced by a function field $L \supset K$. Therefore, $\{x, y\}/\{u, v\}$ and $\{x', y'\}/\{u', v'\}$ are equal in K if and only if they are equal in L, which gives an inclusion of $\mathcal{D}_K$ in $\mathcal{D}_L$. Kronecker believed it was important to structure the theory of divisors in such a way that divisors would be independent of the ambient field in this sense.

§3.11 Numerical extensions.

It is usual in algebraic geometry to consider function fields over an *algebraically closed* field—the field of complex numbers or the field of algebraic numbers—rather than over **Q**. In the Kroneckerian approach, the transfinite construction of algebraically closed fields is avoided by the simple expedient of adjoining new algebraic numbers to **Q** *as needed*.

DEFINITION. A *numerical extension* of a function field K is an extension field $L \supset K$ which is generated over K by a finite number of constants of L.

Clearly, a numerical extension of a function field is a function field, and a numerical extension of a numerical extension is a numerical extension. Moreover, given two numerical extensions L and L' of the same function field K, there is a third numerical extension L'' of K which contains subextensions isomorphic to L and L'.

§3.12 The Idea of a Place.

The notion of primality of divisors depends on the ambient field—an integral divisor prime in a function field K may not be prime in an extension of K. However, there may be divisors prime in K which are prime in all *numerical* extensions of K. Such divisors correspond, roughly, to points on an algebraic curve; more precisely, they correspond to what are called places on an algebraic curve, as will be shown below. Therefore, such divisors in K will be called places in K.

Places will not be *defined* in this way, however. Instead, they are defined in the next article by an algebraic property, and only at a later point in the development (see §3.22) is it shown that places, and only places, are prime divisors in function fields which remain prime in all numerical extensions.

§3.13.

DEFINITION. A *place* in a function field K is an integral divisor in K which can be presented in the form $\{x, y\}$, where $x, y \in K$ are such that (1) K is a numerical extension of the subfield $\mathbf{Q}(x, y)$ of K generated by x and y over $\mathbf{Q}$, and (2) x and y satisfy a polynomial equation $F(x, y) = 0$, with coefficients which are constants of K, in which the constant term is zero but at least one of the two linear terms is nonzero.

Note that the notion of "place", like the notion of "prime", *depends on the ambient field*. However, as is seen immediately from the definition, a place in K is a place in any *numerical* extension of K.

In the language of algebraic geometry, a function field K can be regarded as a field of rational functions on an algebraic curve defined by an irreducible equation of the form $F(x,y) = 0$, where x and y are elements of K such that K is a numerical extension of $\mathbf{Q}(x,y)$. The origin $(x,y) = (0,0)$ is a *non-singular* point of the curve if it is a point of the curve—that is, if the constant term of F is zero—and if the differential of F at (0,0) is not zero—that is, if the linear terms of F are not both zero. Thus, a divisor in K is a place if there is a presentation of K as the field of rational functions on a curve in which the divisor is $\{x,y\}$ and in which the origin is a nonsingular point of the curve.

Example: Let $K = \mathbf{Q}(x,y)$ be the extension of $\mathbf{Q}(x)$ obtained by adjoining a root y of $x^3 + Y^3 - xY$. (Otherwise stated, $K = V_P$—see §1.23—where P is the prime* divisor in $\mathbf{Q}(X,Y)$ represented by $X^3 + Y^3 - XY \in \mathbf{Q}[X,Y]$.) Geometrically, the line $x = y$ intersects the "folium of Descartes" $x^3 + y^3 = xy$ at two points, $(0,0)$ and $(\frac{1}{2}, \frac{1}{2})$. The latter point is a nonsingular point of the folium, because $d(x^3 + y^3 - xy) = 3x^2\,dx + 3y^2\,dy - x\,dy - y\,dx$, which at $(\frac{1}{2}, \frac{1}{2})$ is $\frac{1}{4}dx + \frac{1}{4}dy \neq 0$. Therefore, $\{x - \frac{1}{2}, y - \frac{1}{2}\}$ is a place in K. The origin, however, is a singular point of the folium, one at which, geometrically, the curve has *two coincident* points. This statement about the geometry of the curve is mirrored by the algebraic fact that *the divisor $\{x,y\}$ in K is a product of two places*, which can be shown as follows.

$$x^3 + y^3 = xy$$

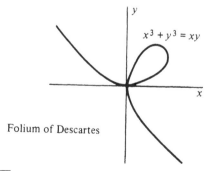

Folium of Descartes

---

*$X^3 + Y^3 - XY$ is irreducible over $\mathbf{Q}(X)$ because it has no root in $\mathbf{Q}[X]$.

The geometry of the curve (see figure) suggests that $m = y/x$ is a "function" on the curve which has two different values, 0 and ∞, at the origin. Substitution of $y = mx$ in the defining equation $x^3 + y^3 - xy = 0$ and cancellation of x^2 (which is nonzero in K) gives $x + m^3x - m = 0$, $x = m/(1 + m^3)$. The restriction to m of $\{x\}$ is the numerator of $[m]_m/[1 + m^3]_m$, which, because numerator and denominator in this quotient are relatively prime, is $[m]_m$. Similarly, the restriction of $\{y\} = \{mx\} = \{m^2/(1 + m^3)\}$ to m is $[m]_m^2$. Therefore, the restriction to m of $\{x, y\}$ is $[m]_m$. Since $1/m = x/y$, and the defining equation is symmetric in x and y, it follows without further calculation that the restriction of $\{x, y\}$ to $1/m$ is $[1/m]_{1/m}$. These two restrictions determine $\{x, y\}$. The restriction of $\{1/m\}$ to m is the numerator of $[1/m]_m$, which is the denominator of $[m]_m$, which is $[1]_m$. Thus, $\{x, y\}$ and $\{m\}\{1/m\}$ have the same restriction $[m]_m$ to m. Symmetrically, they have the same restriction $[1/m]_{1/m}$ to $1/m$. Therefore, $\{x, y\} = \{m\}\{1/m\}$. The factors on the right are places in K because $K = \mathbf{Q}(m)$ (both x and y were shown above to be in $\mathbf{Q}(m)$) and the equations $m - m = 0$ and $(1/m) - (1/m) = 0$ show that $\{m\} = \{m, m\}$ and $\{1/m\} = \{1/m, 1/m\}$ are places of $\mathbf{Q}(m)$.

§3.14.

The definitions of §1.23 have obvious adaptations to divisors in function fields: An element z of a function field K is *finite* at an integral divisor A in K if z is finite at A_x for all parameters x of K; similarly, z is *zero* at A if z is zero at A_x for all parameters x. An element zero at A is also finite at A, and 0 is zero at A for all A. For $z \neq 0$, $[z] = \{z\}/\{1/z\}$, and z is zero at A if and only if $A|\{z\}$, finite at A if and only if $[\{1/z\}, A] = [1]$.

The elements of K finite at A are a *ring*, and the elements of K zero at A are an *ideal* of this ring (because elements finite at A_x are a ring, and elements zero at A_x an ideal of this ring, for all x). The quotient ring is the *ring of values* of K at A, denoted V_A. Since a constant of K is zero at A if and only if it is 0, V_A contains the field of constants of K in a natural way. Therefore, V_A is a vector space over the field of constants of K. The theorem of the next article implies that the constants of K account for all of V_A when A is a place, that is, every element of K finite at a place can be assigned a constant "value" at that place.

§3.15 Local Parameters at a Place.

THEOREM. *Let P be a place in K. There is a unique homomorphism ord_P from the multiplicative group K^* of nonzero elements of K onto the additive group $\mathbf{Z}$ such that (1) $z \in K^*$ is finite at P if and only if $\mathrm{ord}_P(z) \geq 0$, and (2) $z \in K^*$ is zero at P if and only if $\mathrm{ord}_P(z) > 0$. If $z \in K^*$ and $\mathrm{ord}_P(z) = 0$, there is a unique nonzero constant a of K such that $z - a$ is zero at P.*

PROOF: Let $P = \{x, y\}$ be a presentation of P as a place, and let n be the degree of the highest order terms of the polynomial $F(X, Y)$ in the relation $F(x, y) = 0$ which x and y are assumed to satisfy. Let a, b, c, d be rational numbers to be determined later, and let $x' = ax + by$, $y' = cx + dy$. Then x' and y' satisfy $G(x', y') = 0$, where $G(X, Y) = F((dX - bY)/(ad - bc), (-cX + aY)/(ad - bc))$. The coefficients of X^n and Y^n in $G(X, Y)$ are nonzero polynomials in a, b, c, d divided by $(ad - bc)^n$, and there are no terms in $G(X, Y)$ of degree greater than n. Choose a, b, c, d to make these coefficients nonzero and to make $ad \neq bc$. Since $\mathbf{Q}(x, y) = \mathbf{Q}(x', y')$ (by assumption, $ad - bc \neq 0$, so the expression of x' and y' in terms of x and y is invertible), since the constant term of G is zero, and since the linear terms of G are not both zero (because the linear terms of F are not both zero), it follows that one can assume without loss of generality that the polynomial relation satisfied by x and y contains x^n and y^n with nonzero coefficients, and contains no terms of degree greater than n. In this case, y is integral over $\mathbf{Q}[x]$ and x is integral over $\mathbf{Q}[y]$.

By assumption, K is a numerical extension of $\mathbf{Q}(x, y)$, and the coefficients of F are constants of K. Therefore, there is an algebraic number field $K_0 \subset K$ which contains all the coefficients of F and which satisfies $K = K_0(x, y)$. Since x and y can be interchanged if necessary, one can assume without loss of generality that the coefficient of y in F is nonzero, and in fact, since one can divide by this coefficient, that it is 1. Let all terms of the relation $F(x, y) = 0$ which contain x be moved to the right side of the equation, and let a factor of y be taken out on the left side and a factor of x be taken out on the right, to put the relation in the form $y\phi(y) = x\psi(x, y)$, where ϕ and ψ are polynomials with coefficients in K_0 and $\phi(0) = 1$.

Every nonzero element z of K can be expressed in the form

$$z = \frac{\theta(x, y)}{\eta(x, y)},$$

where numerator and denominator are nonzero elements of K expressed as polynomials in x and y with coefficients in K_0. If $\theta(0, 0) = 0$, then $\theta(x, y) = x\sigma(x, y) + y\tau(x, y)$ and $\theta(x, y)\phi(y) = x\theta_1(x, y)$, where the polynomial θ_1 is defined by $\theta_1(X, Y) = \sigma(X, Y)\phi(Y) + \psi(X, Y)\tau(X, Y)$. If $\theta_1(0, 0) = 0$, this process can be repeated to give $\theta(x, y)\phi(y)^2 = x^2\theta_2(x, y)$. After a finite number of steps, one must reach an equation of the form $\theta(x, y)\phi(y)^\mu = x^\mu\theta_\mu(x, y)$ in which $\theta_\mu(0, 0) \neq 0$, because, since $[x^\mu\theta_\mu(x, y)]_x$ is divisible at least μ times by $P_x = [x, y]_x$ and $[\phi(y)]_x$ is not divisible at all by P_x ($\phi(0) = 1$), μ cannot exceed the number of times P_x divides $[\theta(x, y)]_x$. Therefore, multiplication of numerator and denominator of $z = \theta/\eta$ by a sufficiently high power of $\phi(y)$ gives

$$z = \frac{x^\mu\theta_\mu(x, y)\phi(y)^{N-\mu}}{x^m\eta_m(x, y)\phi(y)^{N-m}},$$

where $\theta_\mu(0,0)$ and $\eta_m(0,0)$ are nonzero. With $j = \mu - m$, and $a = \theta_\mu(0,0)/\eta_m(0,0)$, this relation can be written in the form

$$\frac{z}{x^j} - a = \frac{\theta_\mu(x,y)\phi(y)^{N-\mu}}{\eta_m(x,y)\phi(y)^{N-m}} - \frac{\theta_\mu(0,0)}{\eta_m(0,0)} = \frac{\delta(x,y)}{c + \varepsilon(x,y)},$$

where $j \in \mathbf{Z}$, where $a, c \in K_0$ are nonzero, and $\delta(x,y)$ and $\varepsilon(x,y)$ are polynomials in x and y with constant term 0.

The restriction of $\{zx^{-j} - a\}$ to x is therefore the numerator of the restriction to x of $[\delta(x,y)]/[c + \varepsilon(x,y)]$. Since y is integral over $\mathbf{Q}[x]$, both $[\delta(x,y)]_x$ and $[c + \varepsilon(x,y)]_x$ are integral. Since P_x divides the former and is relatively prime to the latter, $P_x|\{zx^{-j} - a\}_x$. Since x supports P ($P|\{x\}$ and $\{x\}$ is relatively prime to $\{1/x\}$), $P|\{zx^{-j} - a\}$, that is, $zx^{-j} - a$ is zero at P.

Given z, j is characterized by the property that $zx^{-j} = a + (zx^{-j} - a)$ is finite and nonzero at P, because, if $i < j$, then $zx^{-i} = x^{j-i}zx^{-j}$ is zero at P, and, if $i > j$, then zx^{-i} must not be finite at P, because, if it were, then $zx^{-j} = x^{i-j}zx^{-i}$ would be zero at P. Therefore, the function $z \mapsto j$ is a *homomorphism* from K^* to $\mathbf{Z}$, because if zx^{-j} and wx^{-i} are finite and nonzero at P the same is true of their product zwx^{-j-i}. This homomorphism carries x to 1. If it carries z to $j \geq 0$, then $z = x^j(zx^{-j} - a) + ax^j$ is finite at P, and if it carries z to $j > 0$, then z is zero at P (elements of K zero at P are an ideal in the ring of elements of K finite at P). Thus, $\text{ord}_P(z) = j$ defines a homomorphism $K^* \to \mathbf{Z}$ with the required properties.

If ord_P is a homomorphism with the required properties, then $\text{ord}_P(zx^{-j})$ is nonnegative but not positive; therefore, $\text{ord}_P(zx^{-j}) = 0$, that is, $\text{ord}_P(z) = j\,\text{ord}_P(x)$. Because ord_P is onto, $\text{ord}_P(x)$ (which is positive) must be 1. Thus, $\text{ord}_P(z) = j$ is the only such homomorphism.

Finally, when $j = 0$, the above construction of an element $zx^{-j} - a$ zero at P gives $z - a$ zero at P. If $z - b$ is also zero at P, where b is a constant of K, then $b - a = (z - a) - (z - b)$ is a constant of K which is zero at P, that is, $b = a$.

DEFINITION. An element $x \in K^*$ for which $\text{ord}_P(x) = 0$ is called a *local parameter* of K at the place P.

§3.16.

COROLLARIES. (1) *A place in a function field K is prime in K and in any numerical extension of K.* (2) *Let x be a local parameter at a place P in a function field K. If z is any nonzero element of K and if $\nu = \text{ord}_P(z)$, there is an infinite sequence $a_\nu, a_{\nu+1}, a_{\nu+2}, \ldots$ of constants of K with $a_\nu \neq 0$, such that, for all $N \geq \nu$, $z = a_\nu x^\nu + \cdots + a_N x^N + o(x^N)$, that is, $(z - a_\nu x^\nu - a_{\nu+1} x^{\nu+1} - \cdots - a_N x^N) x^{-N}$ is zero at P.*

DEDUCTIONS: (1) Let A be an integral divisor in K such that $P \nmid A$. Then $B = P/[P, A]$ is integral, $B|P$, and $B \neq [1]$. Let $u, v \in K$ be such that $B = \{u, v\}$. Since $B \neq [1]$ and B divides both $\{u\}$ and P, $[P, \{u\}] \neq [1]$, which is to say that $1/u$ is not finite at P. Thus, $\text{ord}_P(1/u) < 0$, $\text{ord}_P(u) > 0$, so u is zero at P, that is, $P|\{u\}$. For the same reason, $P|\{v\}$. Thus, $P|\{u, v\} = B$, which implies $P = B = P/[P, A]$, that is, $[P, A] = [1]$. Since $P \nmid A$ implies $[P, A] = [1]$, P is prime.

(2) For $N = \nu, \nu + 1, \nu + 2, \ldots$, let $q_N = (z - a_\nu x^\nu - a_{\nu+1} x^{\nu+1} - \cdots - a_N x^N)/x^N$, where the a_i are constants in K to be determined. The theorem provides the unique nonzero a_ν such that $q_\nu = zx^{-\nu} - a_\nu$ is zero at P. Then $\text{ord}_P(q_\nu) > 0$, $\text{ord}_P(q_\nu/x) \geq 0$, and the theorem provides a constant $a_{\nu+1}$ (zero if q_ν/x is zero at P) such that $q_{\nu+1} = (q_\nu/x) - a_{\nu+1}$ is zero at P. In the same way, once $a_\nu, a_{\nu+1}, \ldots, a_{N-1}$ have been found, a_N is determined by the condition that $q_N = (q_{N-1}/x) - a_N$ be zero at P.

It is a corollary to the *proof* of the theorem that if a place in a function field K is given, then the field of *all* constants in K can be constructed. All constants in the computation of a, starting with the coefficients of ϕ, ψ, θ, η, are in the algebraic

number field K_0, so a is in K_0. In particular, if $z \in K$ is
constant, then $P|\{z - a\}$ for some $a \in K_0$. Since $z - a$ is
constant, $z - a = 0$, $z = a \in K_0$, and K_0 is the field of all
constants of K, as was to be shown.

§3.17 Relative norms.

For the sake of simplicity, only *absolute* norms of divisors
were defined in §1.17. In the case of function fields, in which
the divisor theory is defined relative to a whole family of natural rings $\mathbf{Q}[x]$ and there is no one field with respect to which
"absolute" norms can be defined, it is natural to introduce *relative* norms, which requires only a slight modification of the
proposition of §1.17.

PROPOSITION. *Let K and L be extensions of a natural ring r
of finite degree, and let $L \supset K$. Given a polynomial f with
coefficients in L, let $N_{L/K}f$ be the polynomial with coefficients
in K obtained by taking the determinant of the matrix which
represents multiplication by f relative to a basis of L over K.
(Clearly $N_{L/K}f$ is independent of the choice of the basis.) If
the coefficients of f are integral over r, then $N_{L/K}f = fq$,
where the coefficients of q are elements of L integral over r.*

PROOF: There exist elements ω_1, ω_2, ..., ω_μ of L such that
any coefficient of f times any ω is a linear combination of ω's
with coefficients in r (Proposition 1 of Part 0). Then $\omega f =
M\omega$, where ω is the column vector with entries $\omega_1, \omega_2, \ldots, \omega_\mu$,
where f denotes the 1×1 matrix, and where M is a $\mu \times \mu$ matrix
whose entries are polynomials in the indeterminates of f with
coefficients in r. In other words, $(fI - M)\omega = 0$, where I is the
$\mu \times \mu$ identity matrix and 0 is the column matrix containing
μ zeros. Since ω is nonzero, it follows that $\det(fI - M) = 0$,
that is, f is a root of $\det(XI - M) = F(X) = X^\mu + F_1 x^{\mu-1} +
F_2 x^{\mu-2} + \cdots + F_\mu$, where X is a new indeterminate and the
F_i are polynomials in the indeterminates of f with coefficients
in r. By the Remainder Theorem, $F(X) = (X - f)Q(X)$,

where the coefficients of Q are polynomials with coefficients in L. Taking $N_{L/K}$ on both sides gives $F(X)^n = N_{L/K}(X - f)N_{L/K}Q(X)$ because the norm of a product is clearly the product of the norms, and because the coefficients of F are in $r \subset K$. (Here $n = [L:K]$.) Since $F(X)^n$ has coefficients in r and $N_{L/K}Q(X)$ is monic, the coefficients of $N_{L/K}(X - f)$ are roots of polynomials whose coefficients are polynomials in the coefficients of $F(X)^n$ with coefficients in $\mathbf{Z}$ (by the theorem of Part 0 and the method of Corollary 3 of Part 0 for reducing from the case of several indeterminates to the case of one). Therefore, the coefficients of $N_{L/K}(X - f)$ are integral over r (Proposition 2 of Part 0).

Say $N_{L/K}(X - f) = X^n + A_1 X^{n-1} + \cdots + A_n$, where the A's are polynomials (in the indeterminates of f) with coefficients in K integral over r. Substitution of $X = 0$ gives $(-1)^n N_{L/K} f = A_n$. Substitution of $X = f$ in the matrix of which $N_{L/K}(X-f)$ is the determinant gives a matrix whose determinant is zero (the chosen basis of L over K is a column vector in its kernel). Therefore $f^n + A_1 f^{n-1} + \cdots + A_n = 0$, which gives $N_{L/K} f = \pm A_n = \mp(f^n + A_1 f^{n-1} + \cdots + A_{n-1} f)$, and proves the proposition with $q = \mp(f^{n-1} + A_1 f^{n-2} + \cdots + A_{n-1})$.

COROLLARY. *If K and L are function fields with $K \subset L$, and if A is a global divisor in L, then $N_{L/K} A$ is a global divisor in K, defined in the obvious way: For any parameter x of K, let f be a polynomial with coefficients in L such that $A_x = [f]_x$ and set $(N_{L/K} A)_x = [N_{L/K} f]_x$. If A is integral, then $A | N_{L/K} A$. In particular, $N_{L/K} A$ is integral if A is.*

DEDUCTION: By the proposition, with $r = \mathbf{Q}[x]$, if $[f]_x$ is integral, then $[f]_x | [N_{L/K} f]_x$, and, in particular, $[N_{L/K} f]_x$ is integral. It follows, as in §1.17, that $[f]_x | [g]_x$ implies $[N_{L/K} f]_x | [N_{L/K} g]_x$. Therefore, $(N_{L/K} A)_x$ is independent of the choice of f. Moreover, if x is integral over $\mathbf{Q}[u]$ and $A_x = [f]_x$, then $A_u = [f]_u$, which shows that $(N_{L/K} A)_u = [N_{L/K} f]_u$ is the restriction to u of $(N_{L/K} A)_x = [N_{L/K} f]_x$, so $N_{L/K} A$ is a well-defined divisor. If A is integral, $A | N_{L/K} A$ because $A_x | N_{L/K} A_x$ for all x.

Note: The homomorphism $N_{L/K} : \mathcal{D}_L \to \mathcal{D}_K$ *depends on the ambient field* L, as well as the field K to which the norm is taken. For example, if A is a divisor in L, and if $L' \supset L$, then $N_{L'/K} A = (N_{L/K} A)^{[L':L]}$.

§3.18 A Fundamental theorem.

THEOREM. *Given a divisor A in a function field K, there is a numerical extension of K in which A is a product of powers (possibly negative) of places.*

(Since places are prime, the representation of a divisor as a product of powers of places in a numerical extension is *unique* in a natural way, namely: Let L and L' be numerical extensions of K in which A is a product of powers of places, say $A = \prod_i P_i^{\nu_i}$ in L and $A = \prod_i Q_i^{\mu_i}$ in L'. There is a numerical extension L'' of K in which L and L' can both be embedded isomorphically. Under such isomorphisms, both the P_i and the Q_i correspond to places in L''. By the uniqueness of factorization of divisors in L'', the two factorizations of A must be identical in L''.)

LEMMA. *Given an integral divisor A in a function field K, there is an upper bound on the number of nontrivial factors in a factorization of A in a numerical extension of K. More precisely, there is an integer ν such that if $A = B^{(1)} B^{(2)} \cdots B^{(\mu)}$, where the $B^{(i)}$ are integral divisors in a numerical extension of K and where $\mu > \nu$, then $B^{(i)} = [1]$ for at least one i.*

PROOF: Let $x \in K$ be such that $A | \{x\}$. It will be shown that $\nu = [K : \mathbf{Q}(x)]$ has the desired property.

Let $A = \prod_{i=1}^{\mu} B^{(i)}$, where the $B^{(i)}$ are integral divisors in a numerical extension L of K. Let $L = K(c_1, c_2, \ldots, c_\rho)$, where the c_i are constants of L. Let $L_1 = \mathbf{Q}(c_1, c_2, \ldots, c_\rho)$, an algebraic number field contained in L. Then $L_1(x)$ is a function field contained in L. Let N denote the norm $N_{L/L_1(x)}$.

Since $A|\{x\}$, $NA|N\{x\} = \{x\}^{[L:L_1(x)]}$. Since $\{x\}$ is a place in the function field $L_1(x)$ and is therefore prime in $L_1(x)$, NA is thus a power of $\{x\}$, say $NA = \{x\}^\tau$, where τ satisfies $0 \leq \tau \leq [L:L_1(x)]$. In the same way, $NB^{(i)} = \{x\}^{\sigma_i}$, for some $\sigma_i \geq 0$, for each $i = 1, 2, \ldots, \mu$. Clearly, $\sigma_1 + \sigma_2 + \cdots + \sigma_\mu = \tau$. If $\sigma_i = 0$, then $B^{(i)}|\{x\}^0 = [1]$, and $B^{(i)} = [1]$. Therefore, if none of the factors of $A = \prod B^{(i)}$ is $[1]$, each σ_i is at least 1, and $\tau = \sigma_1 + \sigma_2 + \cdots + \sigma_\mu \geq \mu$. Since $\tau \leq [L:L_1(x)]$, it follows that $[L:L_1(x)] \geq \mu$.

A basis of $L_1(x)$ over $\mathbf{Q}(x)$ spans L_1 over $\mathbf{Q}(x)$. Since L_1 spans L over $K \supset \mathbf{Q}(x)$, a basis of $L_1(x)$ over $\mathbf{Q}(x)$ spans L over K. Therefore, $[L:K] \leq [L_1(x):\mathbf{Q}(x)]$. Since $[L:K][K:\mathbf{Q}(x)] = [L:\mathbf{Q}(x)] = [L:L_1(x)][L_1(x):\mathbf{Q}(x)]$, it follows that $\nu = [K:\mathbf{Q}(x)] \geq [L:L_1(x)] \geq \mu$ if no $B^{(i)}$ is $[1]$, as was to be shown.

PROOF OF THE THEOREM: Since every divisor is a quotient of integral divisors, it will suffice to prove the theorem for integral divisors A. If $A = [1]$, the theorem is true by virtue of the convention that $[1]$ is a product of *no* places. Assume, therefore, that A is integral and not $[1]$. The method of proof will be to give a construction which *either* produces a proper factorization of A in a numerical extension of K *or* proves that A is a place in a numerical extension of K. Since, by the lemma, the number of times A can be factored is bounded, repeated application of this construction will succeed eventually in writing A as a product of places in a numerical extension of K.

Let x be a parameter of K such that $A^{-1}\{x\}$ is integral and relatively prime to A (§3.10). By the theorem of the primitive element (§1.25), the extension $K \supset \mathbf{Q}(x)$ is generated by a single element $y \in K$. Let $F \in \mathbf{Q}[X,Y]$ be an irreducible polynomial such that $F(x,y) = 0$ in K. Then $K = \mathbf{Q}(x,y)$ is obtained by adjoining to $\mathbf{Q}(x)$ a root y of $F(x,Y) = 0$.

By the proposition of §3.6, $A = BC$, where B is supported by y and C is concentrated at $\{1/y\}$. If neither B nor C is [1], the construction terminates with a proper factorization $A = BC$ of A. If $B = [1]$, $A = C$ is concentrated at $\{1/y\}$ and is therefore supported by $1/y$. If $C = [1]$, $A = B$ is supported by y. In either case, then, there is a parameter of K which generates K over $\mathbf{Q}(x)$ and which supports A. Let y be such a parameter.

The restriction to y of $N_{K/\mathbf{Q}(y)}A$ is integral, call it $[q(y)]_y$, where $q(y)$ is a polynomial with rational coefficients. There is a numerical extension L of K in which $q(y)$ splits into linear factors, say $[q(y)]_y = \prod[y - k_i]_y$, where the k_i are constants in L. (A splitting field of q over $\mathbf{Q}$ can be obtained by adjoining a single root of a single irreducible polynomial with coefficients in $\mathbf{Q}$. Such a polynomial has coefficients in K, and adjunction of a single root of a single factor of it irreducible over K gives a field L with the desired property.) If one of the divisors $[A, \{y - k_i\}]$ is a proper divisor of A, then A is factored and the construction terminates. Otherwise, every $[A, \{y - k_i\}]$ is [1] or A. If $[A, \{y - k_i\}]$ were [1] for all roots k_i of q, then $[A_y, y - k_i]_y$ would be $[1]_y$ for all k_i, which would imply that A_y was relatively prime to $[q(y)]_y = \prod[y - k_i]_y$; since A_y divides its norm $[q(y)]_y$, this would imply $A_y = [1]_y$, contrary to the assumptions that $A \neq [1]$ and y supports A. Therefore, $A|\{y - k\}$ for some constant k of L. (In fact, since $[\{y - k\}, \{y - k'\}] = [1]$ for $k \neq k'$, this condition determines k.) In summary, if K is a numerical extension of $\mathbf{Q}(x, y)$, and if y supports A, then either A has a proper factor of the form $[A, \{y - k\}]$ for some constant k in a numerical extension of K, or A divides $\{x, y - k\}$ for some constant k in a numerical extension of K.

In the second case, if $\{x, y - k\}$ is a place of $L = K(k)$, then, because places are prime and $A \neq [1]$, $A = \{x, y - k\}$ is a place in L and the construction terminates. Otherwise, the construction continues as follows. Let $y_1 = (y - k)/x$. Then L is a numerical extension of $\mathbf{Q}(x, y_1) \subset L$ because adjunction of the constant k to $\mathbf{Q}(x, y_1)$ gives a field containing $y = k + xy_1$, and therefore gives a field containing $\mathbf{Q}(x, y)$, of which L is a numerical extension. Also, y_1 supports A because

$$[y_1] = [y - k]/[x] = \frac{\{y - k\}\{1/x\}}{\{1/(y - k)\}\{x\}}$$

and when the common factor A in numerator and denominator of this quotient is cancelled the denominator becomes $\{1/(y - k)\}A^{-1}\{x\}$, which is relatively prime to A (because $\{1/(y-k)\}$ is relatively prime to $\{y - k\}$ and therefore to A, whereas $A^{-1}\{x\}$ is relatively prime to A by the choice of x). As before, either A has a proper divisor of the form $[A, \{y_1 - k'\}]$ where k' is a constant in a numerical extension of K, or A divides $\{x, y_1 - k'\}$, where k' is a constant in a numerical extension of K. In the first case, the construction terminates. In the second case, the construction terminates if $\{x, y_1 - k'\}$ is a place, but if $\{x, y_1 - k'\}$ is not a place the construction continues by setting $y_2 = (y_1 - k')/x$ and iterating.

If, after m iterations, the construction has not terminated, it has produced a sequence $k, k', k'', \ldots, k^{(m-1)}$ of constants in a numerical extension of K, say in $L = K(k, k', \ldots, k^{(m-1)})$, and a sequence $y, y_1, y_2, \ldots, y_m$ of parameters of L such that $A | \{x, y_i - k^{(i)}\}$ for $i = 0, 1, \ldots, m - 1$, such that $y_{i+1} = (y_i - k^{(i)})/x$ for $i = 0, 1, \ldots, m-1$, and such that $\{x, y_i - k^{(i)}\}$ is *not* a place in L. The theorem will be proved if it is shown that these conditions imply an upper bound on m.

As above, let $F \in \mathbf{Q}[X, Y]$ be an irreducible polynomial with rational coefficients such that $F(x, y) = 0$ in K. Since $\{x, y - k\}$ is not [1] and not a place in L, when F is written as a polynomial in X and $Y - k$, it contains no terms of degree less than 2. Substitution of XY for $Y - k$ in this polynomial therefore gives a polynomial in which all terms are divisible by X^2, say $F(X, k + XY) = X^2 G_1(X, Y)$. Then $G_1(x, y_1) = x^{-2}F(x, k + xy_1) = x^{-2}F(x, y) = 0$ in K, and therefore in L. In the same way, since $\{x, y_1 - k'\}$ is not [1] and not a place,

$G_1(X, k' + XY) = X^2 G_2(X, Y)$, where G_2 is a polynomial whose coefficients are constants of L and where $G_2(x, y_2) = 0$. In general, $G_i(X, k^{(i)} + XY) = X^2 G_{i+1}(X, Y)$ for $i = 1, 2, \ldots, m$. Therefore, $F(X, k + k'X + k''X^2 + \cdots + k^{(m-1)} X^{m-1} + X^m Y) = X^2 G_1(X, k' + k''X + \cdots + k^{(m-1)} X^{m-2} + X^{m-1} Y) = X^4 G_2(X, k'' + k^{(3)} X + \cdots + k^{(m-1)} X^{m-3} + X^{m-2} Y) = \cdots = X^{2m-2} G_{m-1}(X, k^{(m-1)} + XY) = X^{2m} G_m(X, Y)$. In particular, X^{2m} divides $F(X, k + k'X + k''X^2 + \cdots + k^{(m-1)} X^{m-1} + X^m Y)$ (a polynomial in X and Y with coefficients in the algebraic number field $\mathbf{Q}(k, k', \ldots, k^{(m-1)})$).

This condition implies an upper bound on m as follows. Let $F_Y(X, Y)$ denote the partial derivative of F with respect to Y (that is, the coefficient of h in $F(X, Y + h)$). When F and F_Y are regarded as polynomials in Y with coefficients in the field $\mathbf{Q}(X)$ of rational functions of X, the Euclidean algorithm can be used to write their greatest common divisor $d(Y)$ in $\mathbf{Q}(X)[Y]$ in the form $a(Y)F + b(Y)F_Y$, where a and b have coefficients in $\mathbf{Q}(X)$. The statement that $d(Y)$ divides F—say $c(Y)d(Y) = F$—implies $c(X, Y)d(X, Y) = e(X)F(X, Y)$ when $e(X) \in \mathbf{Q}[X]$ clears the denominators in c and d to make them polynomials $c(X, Y)$, $d(X, Y) \in \mathbf{Q}[X, Y]$. By unique factorization of polynomials in $\mathbf{Q}[X, Y]$, and by the irreducibility of $F(X, Y)$ in $\mathbf{Q}[X, Y]$, F divides either $c(X, Y)$ or $d(X, Y)$ and the other is a polynomial in X alone. Since $d(Y)$ divides F_Y, its degree in Y is at most $\deg_Y(F) - 1$, which shows that F cannot divide $d(X, Y)$. Therefore, $d(X, Y)$ is a polynomial in X alone of the form $f(X)a(Y)F + f(X)b(Y)F_Y$, say $\Delta(X) = g(X, Y)F(X, Y) + h(X, Y)F_Y(X, Y)$, where g, $h \in \mathbf{Q}[X, Y]$. In this equation, let Y be replaced by $k + k'X + k''X^2 + \cdots + k^{(m-1)} X^{m-1} + X^m Y$. The left side does not change. The first term on the right is divisible by X^{2m}. The second term on the right is divisible by X^m because differentiation of $F(X, k + k'X + \cdots + X^m Y) = X^{2m} G_m(X, Y)$ with respect to Y gives $X^m F_Y(X, k + k'X + \cdots + X^m Y) = X^{2m}$ times a polynomial in X and Y. Therefore, X^m must divide

$\Delta(X)$. Since $\Delta(X)$ can be computed *a priori*, knowing only $F(X, Y)$, this condition bounds m and proves the theorem.

§3.19.

COROLLARY. *The field of constants of a function field can be constructed.*

DEDUCTION: Given a function field K, let x be a parameter of K. Since $\{x\} \neq [1]$, the factorization of $\{x\}$ into places in a numerical extension L of K provides a place P in L. As was noted at the end of §3.16, knowledge of P makes possible the construction of the field of constants L_0 of L. Let $1, \alpha_2, \alpha_3, \ldots,$ $\alpha_\nu \in L_0$ be a basis of L over K. Each $\beta \in L_0$ can be written in just one way in the form $\beta = g_1(\beta) + g_2(\beta)\alpha_2 + \cdots + g_\nu(\beta)\alpha_\nu$, and this equation defines functions $g_i : L_0 \to K$ linear over $\mathbf{Q}$. As a vector space over $\mathbf{Q}$, the field of constants K_0 of K is the subspace of L_0 defined by the relations $g_i(\beta) = 0$ for $i = 2, 3,$ $\ldots, \nu$. Thus, a basis of K_0 over $\mathbf{Q}$ can be constructed.

§3.20 Presentation of Places.

THEOREM. *Let x and y be elements of a function field K such that $\{x, y\}$ is a place in K. Then x and y present $\{x, y\}$ as a place in K, that is, (1) K is a numerical extension of $\mathbf{Q}(x, y)$ and (2) x and y satisfy a polynomial relation $F(x, y) = 0$, with coefficients which are constants of K, in which the constant term is zero but at least one of the two linear terms is nonzero.*

LEMMA. *Let K be a function field, let K_0 be its field of constants, let x, $y \in K$ generate K over K_0, and let $F(X, Y) \in K_0[X, Y]$ be a nonzero polynomial, irreducible over K_0, such that $F(x, y) = 0$ in K. Finally, let $A = \{x, y\}$. Then $N_{K/K_0(x)}A = \{x\}^\lambda$, where λ is the degree of the term or terms of $F(X, Y)$ of lowest degree.*

PROOF: Let $F(X, Y) = a_0(X)Y^m + a_1(X)Y^{m-1} + \cdots + a_m(X)$. Multiplication of the relation $F(x, y) = 0$ by $a_0(x)^{m-1}$ shows that $a_0(x)y$ is integral over $K_0[x]$. Therefore,

$$[x, y]_x = \frac{[a_0(x)x, a_0(x)y]_x}{[a_0(x)]_x}$$

presents $[x, y]_x$ as a quotient of integral divisors. Thus, A_x, which is the numerator of $[x, y]_x$, is $[a_0(x)x, a_0(x)y]_x$ divided by its g. c. d. with $[a_0(x)]_x$, that is, divided by $[a_0(x)x, a_0(x)y, a_0(x)]_x = [a_0(x)y, a_0(x)]_x$. Therefore,

$$NA_x = \frac{N[a_0(x)x, a_0(x)y]_x}{N[a_0(x), a_0(x)y]_x},$$

where N is the norm from K to $K_0(x)$.

Now $[a_0(x)x, a_0(x)y]_x = [a_0(x)]_x[xU - y]_x$, where U is an indeterminate. For indeterminate V, an elementary computation gives

$$N_{K/K_0(x)}(V - y) = V^m + \frac{a_1(x)}{a_0(x)}V^{m-1} + \cdots + \frac{a_m(x)}{a_0(x)}.$$

Therefore,

$N[a_0(x)x, a_0(x)y]_x$

$$= [a_0(x)^m]_x[(xU)^m + \frac{a_1(x)}{a_0(x)}(xU)^{m-1} + \cdots + \frac{a_m(x)}{a_0(x)}]_x$$

$$= [a_0(x)^{m-1}]_x[a_0(x)x^m U^m + a_1(x)x^{m-1}U^{m-1} + \cdots + a_m(x)]_x$$

$$= [a_0(x)^{m-1}]_x[a_0(x)x^m, a_1(x)x^{m-1}, a_2(x)x^{m-2}, \ldots, a_m(x)]_x.$$

The calculation of $N[a_0(x), a_0(x)y]_x$ is the same with $U - y$ in place of $xU - y$, which gives $N[a_0(x), a_0(x)y]_x = [a_0(x)^{m-1}]_x[a_0(x), a_1(x), \ldots, a_m(x)]_x$. The second factor on the right is $[1]_x$ because $F(X, Y)$ is irreducible over K_0, so that the $a_i(x)$ have no common factor in $K_0[x]$. In conclusion, therefore, $NA_x = [a_0(x)x^m, a_1(x)x^{m-1}, \ldots, a_m(x)]_x$.

Since $A|\{x\}$, $NA = \{x\}^\mu$ for some μ ($\{x\}$ is a place in $K_0(x)$). In fact, μ is the largest integer such that $[x]_x^\mu$ divides NA_x, that is, the largest integer such that $[x]_x^\mu$ divides $[a_i(x)x^{m-i}]_x$ for $i = 0, 1, \ldots, m$. If $F(X, Y) = \sum\sum b_{\sigma\tau}X^\sigma Y^\tau$, then $a_i(X) = \sum b_{\sigma,m-i}X^\sigma$, and $a_i(x)x^{m-i} = \sum_\sigma b_{\sigma,m-i}x^{\sigma+m-i}$ is divisible by x^μ if and only if $b_{\sigma,m-i} = 0$ whenever $\sigma + m - i < \mu$. Thus, $[x]_x^\mu$ divides NA_x if and only if $F(X, Y) = \sum\sum b_{\sigma\tau}X^\sigma Y^\tau$ has $b_{\sigma\tau} = 0$ whenever $\sigma + \tau < \mu$. In other words, $[x]_x^\mu$ divides NA_x if and only if $\mu \leq \lambda$, as was to be shown.

PROOF OF THE THEOREM: Let $P = \{x, y\}$ be a place in K. Since either x or y must be a local parameter at P, there is no loss of generality in assuming that x is a local parameter at P. Let $z \in K$ be a generator of K over $K_0(x)$ which is integral over $K_0[x]$, and define a sequence $z_0, z_1, z_2, \ldots$ of elements of K finite at P and a sequence $k_0, k_1, k_2, \ldots$ of elements of K_0 by the conditions $z_0 = z$, $z_i - k_i$ is zero at P, and $z_{i+1} = (z_i - k_i)/x$.

Let $G_0(X, Z)$ be an irreducible polynomial with coefficients in K_0 such that $G_0(x, z_0) = 0$. (K is an algebraic extension of $K_0(x)$.) As a polynomial in x and $z_0 - k_0$, $G_0(x, k_0 + (z_0 - k_0))$ has constant term zero (both x and $z_0 - k_0$ are zero at P, so the constant term is zero at P, which implies that the constant term is zero). If either of its linear terms is nonzero, then, since $K = K_0(x, z) = K_0(x, z_0 - k_0)$, x and $z_0 - k_0$ present P as a place in K. Otherwise, every term of $G_0(X, k_0 + XZ)$ is divisible by X^2 and $G_1(X, Z) = X^{-2}G_0(X, k_0 + XZ)$ is a polynomial in X and Z with coefficients in K_0. Since $G_1(x, z_1) = x^{-2}G_0(x, k_0 + xz_1) = x^{-2}G_0(x, z_0) = 0$, the constant term of $G_1(x, k_1 + (z_1 - k_1))$ as a polynomial in x and $z_1 - k_1$ is zero. If either linear term is nonzero, then, because $K = K_0(x, z_0 - k_0) = K_0(x, z_1 - k_1)$, x and $z_1 - k_1$ present P as a place in K. Otherwise, $G_2(X, Z) = X^{-2}G_1(X, k_1 + XZ)$ is a polynomial with coefficients in K_0 for which $G_2(x, z_2) = 0$, and the process continues. The process must reach a point at which x and $z_i - k_i$ present P as a place in K, because, as was shown in §3.18, the condition that there be a polynomial $G_m(X, Z) = X^{-2}G_{m-1}(X, k_{m-1} + XZ) = X^{-4}G_{m-2}(X, k_{m-2} + k_{m-1}X + X^2 Z) = \cdots = X^{-2m}G_0(X, k_0 + k_1 X + k_2 X^2 + \cdots + k_{m-1}X^{m-1} + X^m Z)$ imposes an upper bound on m. Therefore, there is a $w \in K$ of the form $w = z_i - k_i$ such that x and w present P as a place in K.

Let $K' = K_0(x, y)$. Since $P = \{x, y\}$ is in K',

$$N_{K/K_0(x)}P = \left(N_{K'/K_0(x)}P\right)^{[K:K']}.$$

By the lemma, $N_{K'/K_0(x)}P = \{x\}^\lambda$, where λ is the degree of the term or terms of lowest degree in F. Thus, $N_{K/K_0(x)}P = \{x\}^{\lambda[K:K']}$. On the other hand, since x and w present $P = \{x, w\}$ as a place in K, by the lemma, $N_{K/K_0(x)}P = \{x\}$. Thus, $1 = \lambda[K:K']$, which implies $\lambda = 1$ and $[K:K'] = 1$, that is, F contains a nonzero linear term and $K = K_0(x, y)$, which is a numerical extension of $\mathbf{Q}(x, y)$, as was to be shown.

3.21 The Degree of a Divisor.

The *degree* of a divisor A in a function field K, denoted $\deg_K A$, is the sum of the exponents in a representation of A as a product of powers of places in a numerical extension of K. (The uniqueness of the representation of A as a product of places in a numerical extension of K implies that $\deg_K A$ is well defined. Clearly, when L is a numerical extension of K, $\deg_L A = \deg_K A$. Also, $\deg_K AB = \deg_K A + \deg_K B$.)

THEOREM. *If K and L are function fields with $K \subset L$, there is a positive integer m such that, for all divisors A in K, $\deg_L A = m \deg_K A$. Specifically, $m = [L:K]/[L_0:K_0]$, where L_0 and K_0 are the fields of constants of L and K respectively.*

(Geometrically, the inclusion $K \subset L$ corresponds to a morphism from the curve with field of functions L to the curve with field of functions K. Each place on the range curve corresponds to m places on the domain curve, counted with multiplicities.)

LEMMA 1. *If L is a numerical extension of K, then $[L:K] = [L_0:K_0]$.*

The proof given in the first printing was erroneous. The Lemma on page 119 shows that if β is constant—that is, if β is integral over both $Q[x]$ and $Q[1/x]$—then so are the α_i. Therefore, adjoining a constant to K is the same as adjoining a constant to K_0. (Note added in second printing.)

LEMMA 2. *Let A be an integral divisor in a function field K, and let $A|\{x\}$ for $x \in K$. Then $N_{K/K_0(x)}A = \{x\}^{\deg_K A}$.*

PROOF: If $A = \{x,y\}$ is a place in K, $\deg_K A = 1$ and the lemma follows from the theorem and lemma of §3.20.

Given any integral divisor A in K, let L be a numerical extension of K in which A is a product of places, say $A = P^{(1)}P^{(2)} \cdots P^{(\nu)}$, where the $P^{(i)}$ are places in L and $\nu = \deg_K A$. Let L_0 be the field of constants of L. Then $N_{L/L_0(x)}A = \prod N_{L/L_0(x)}P^{(i)} = \{x\}^\nu$ and what is to be shown is that $N_{L/L_0(x)}A = N_{K/K_0(x)}A$. Since both sides of this equation are powers of $\{x\}$, it will suffice to prove that their restrictions to x are equal, and for this it will suffice to prove that $N_{L/L_0(x)}f = N_{K/K_0(x)}f$ for any polynomial f with coefficients in K. By Lemma 1, $[L:K] = [L_0:K_0] = [L_0(x):K_0(x)]$. Since $[L:K][K:K_0(x)] = [L:K_0(x)] = [L:L_0(x)][L_0(x):K_0(x)]$, it follows that $[K:K_0(x)] = [L:L_0(x)]$. A basis of K over $K_0(x)$ spans both K and L_0 over $L_0(x)$, and therefore spans L over $L_0(x)$. Since $[K:K_0(x)] = [L:L_0(x)]$, it follows that a basis of K over $K_0(x)$ is a basis of L over $L_0(x)$. When such a basis is used to compute $N_{L/L_0(x)}f$ and $N_{K/K_0(x)}f$ for f with coefficients in K, the results are identical, as was to be shown.

LEMMA 3. *If K, L, and M are function fields with $K \subset L \subset M$, then $N_{M/K}A = N_{L/K}(N_{M/L}A)$.*

PROOF: This follows immediately from the same formula for norms of polynomials with coefficients in M, a basic algebraic identity which can be proved by expressing the norm as a product of conjugates under the Galois group (see Proposition 2 of §1.25).

PROOF OF THE THEOREM: Clearly it will suffice to prove that

$$[L_0:K_0] \deg_L A = [L:K] \deg_K A$$

for *integral* divisors A in K. Given such an A, let $x \in K$ satisfy $A|\{x\}$. Then $N_{L/K_0(x)}A = N_{K/K_0(x)}(A^{[L:K]}) = \{x\}^{[L:K]\deg_K A}$ while on the other hand

$$N_{L/K_0(x)}A = N_{L_0(x)/K_0(x)}N_{L/L_0(x)}A$$
$$= N_{L_0(x)/K_0(x)}\{x\}^{\deg_L A} = \{x\}^{[L_0:K_0]\deg_L A},$$

from which the desired equation follows.

COROLLARY. *For any nonzero $x \in K$, $\deg_K[x] = 0$.*

DEDUCTION: If x is a constant of K, then $[x] = [1]$ obviously has degree 0. Otherwise, in $\mathbf{Q}(x) \subset K$, $[x] = \{x\}\{1/x\}^{-1}$ gives $[x]$ as a product of places and shows that its degree in $\mathbf{Q}(x)$ is 0. Therefore, by the theorem, its degree in K is 0.

§3.22 A Characteristic of Places.

PROPOSITION. *An integral divisor A in a function field K which is prime in K and in every numerical extension of K is a place in K.*

PROOF: Let $A = \{x, y\}$ where x, $y \in K$ (§3.10), let K_0 be the field of constants of K, and let $F(X, Y) \in K_0[X, Y]$ be a nonzero polynomial, irreducible over K_0, such that $F(x, y) = 0$. By the lemma of §3.20, $N_{K/K_0(x)}A = (N_{K_0(x,y)/K_0(x)}A)^{[K:K_0(x,y)]} = \{x\}^{\lambda[K:K_0(x,y)]}$, where λ is the degree of the terms of lowest degree in F. On the other hand, $\deg_K A = 1$ by the definition of the degree and the assumption on A. By Lemma 2 of §3.21, $N_{K/K_0(x)}A = \{x\}$. Thus, $\lambda[K:K_0(x, y)] = 1$, which implies both that F contains a nonzero linear term and that K is a numerical extension of $\mathbf{Q}(x, y)$.

§3.23 Dimension of a Divisor.

A divisor A in a function field K will be said to *divide* an element z of K, denoted $A|z$, if, for every parameter x of K, $A_x|z$ (where A_x is the restriction of A to x). Otherwise stated, $A|z$ if $z = 0$ or if $z \neq 0$ and $A|[z]$. If $A|z$ and $z \neq 0$, then $A^{-1}[z]$ is integral, which implies $\deg_K(A^{-1}[z]) \geq 0$, $0 = \deg_K[z] \geq \deg_K(A)$. This observation proves the statement of §3.4 that not every (global) divisor is of the form $[f]$; in fact, a divisor of positive degree cannot even *divide* a divisor of this form (because it would then divide a divisor of the form $[z]$ for $z \neq 0$) much less *be* of this form.

If $A|z$ then $A|az$ for any constant a in K, and if $A|z_1$ and $A|z_2$ then $A|(z_1 + z_2)$. Thus, the set of elements of K divisible by A form a *vector space* over the field of constants of K. The dimension of this vector space will be denoted $\dim_K(A)$. As was just observed, $\deg_K(A) > 0$ implies $\dim_K(A) = 0$. As will be proved in the next article, $\dim_K(A) < \infty$ for all divisors A in K.

PROPOSITION. *If A is a divisor in K, and if L is a numerical extension of K, then $\dim_K(A) = \dim_L(A)$.*

LEMMA. *Let $L = K(a)$ be a numerical extension of K obtained by adjoining a constant a to K. Every element of L can be written in one and only one way in the form $\beta = \alpha_0 + \alpha_1 a + \alpha_2 a^2 + \cdots + \alpha_{n-1} a^{n-1}$, where $n = [L : K]$ and where $\alpha_i \in K$. Given a parameter x of K, β is integral over $\mathbf{Q}[x]$ if and only if each α_i is integral over $\mathbf{Q}[x]$.*

PROOF: Any constant of L is integral over $\mathbf{Q}$ and is therefore integral over $\mathbf{Q}[x] \supset \mathbf{Q}$. Therefore, β is integral over $\mathbf{Q}[x]$ whenever the α_i are all integral over $\mathbf{Q}[x]$. For the proof of the converse, let β be integral over $\mathbf{Q}[x]$. Let $F(X)$ be the monic, irreducible polynomial with (constant) coefficients in K of which a is a root (i.e., $F(X) = N_{L/K}(X - a)$), and let $q(X) = F(X)/(X - a)$. Let $\phi(X)$ be the trace of $\beta q(a)^{-1}q(X)$ relative to the extension $L \supset K$. Then $\phi(X) = \sum \psi^{(i)}(X)$, where the $\psi^{(i)}(X)$ are the n conjugates of $\beta q(a)^{-1}q(X)$ in a splitting field of $F(X)$ over K. Since the coefficients of the $\psi^{(i)}(X)$ are integral over $\mathbf{Q}[x]$ (both β and the coefficients of $q(a)^{-1}q(X)$ are integral over $\mathbf{Q}[x]$), so are the coefficients of $\phi(X)$, say $\phi(X) = \gamma_0 + \gamma_1 X + \gamma_2 X^2 + \cdots + \gamma_{n-1}X^{n-1}$, where the $\gamma_i \in K$ are integral over $\mathbf{Q}[x]$. Thus, $\phi(a) = \gamma_0 + \gamma_1 a + \gamma_2 a^2 + \cdots + \gamma_{n-1}a^{n-1}$. On the other hand, the $\psi^{(i)}(a)$ are all 0 (because a is a root of the conjugates of $q(X)$) except for the one corresponding to the identity conjugation, and that one is $\beta q(a)^{-1}q(a) = \beta$. Therefore, $\beta = \phi(a) = \gamma_0 + \gamma_1 a + \gamma_2 a^2 + \cdots + \gamma_{n-1}a^{n-1}$ where the γ_i are integral over $\mathbf{Q}[x]$. Since the representation of β in this form is unique, the lemma follows.

PROOF OF THE PROPOSITION: There is no loss of generality in assuming that $L = K(a)$ where a is a constant of K (every numerical extension can be obtained as the result of a sequence of such extensions). Then $K_0(a)$ is the field of constants of L (Lemma 1, §3.21), where K_0 is the field of constants of K. Every element β of L can be written in one and only one way in the form $\beta = \alpha_0 + \alpha_1 a + \cdots + \alpha_{n-1} a^{n-1}$, where the α_i are in K. Let x be a given parameter of K. By Corollary (2) of §1.20, there exist elements p and Ψ in K such that an element δ of K (or of any extension of K) is divisible by $A_{\dot{x}}$ if and only if $\delta\Psi/p$ is integral over $\mathbf{Q}[x]$. The lemma then implies, because $\beta\Psi/p = \sum \alpha_i \Psi a^i/p$, that A_x divides β if and only if A_x divides each of the α_i. Since this is true for all parameters x, A divides β if and only if A divides each of the α_i. Thus, if $\mu = \dim_K(A)$ and if $\delta_1, \delta_2, \ldots, \delta_\mu$ is a basis over K_0 of the set of elements of K divisible by A, then the $n\mu$ elements $a^i \delta_j$ are a basis over K_0 of the set S of all elements of L divisible by A, which shows that the dimension of S as a vector space over K_0 is $n\mu = n\dim_K(A)$. The dimension of S as a vector space over L_0 is by definition $\dim_L(A)$, which implies, because $[L_0 : K_0] = n$, that the dimension of S as a vector space over K_0 is $n\dim_L(A)$. Thus $n\dim_K(A) = n\dim_L(A)$, and the proposition follows.

§3.24 The Genus of a Function Field.

THEOREM. *Given a function field K, there is an integer g such that*

$$\dim_K(A) + \deg_K(A) + g > 0$$

for all divisors A in K.

Note that $A = [1]$ gives $1 + 0 + g > 0$, so $g \geq 0$. The *smallest* integer g with the property of the theorem is called the *genus* of K. The proof of the theorem will include a method of finding the genus. It will also include a proof of the statement of §3.23 that $\dim_K(A) < \infty$ for all divisors A.

The formula $\dim_K(A) + \deg_K(A) + g > 0$ will be familiar to many readers in the form $\dim_K(A^{-1}) \geq \deg_K(A) - g + 1$. If A is a product of m places, this formula states that the vector space of "functions" with poles at most at the m places of A (that is, elements z of K such that $A[z]$ is integral or $z = 0$) has dimension at least $m - g + 1$.

PROOF: $\dim_K(A)$ can be found explicitly in the case of the divisor $A = \{1/x\}^{-\nu}$, where ν is a large positive integer and x is a parameter of K found in the following way.

Let K be presented in the form $K = K_0(u, v)$, where K_0 is the field of constants of K, and where u and v are parameters of K which satisfy an irreducible polynomial relation $F(u, v) = 0$ with coefficients in K_0 of degree $n = [K : K_0(u)]$ in v, say

$$(1) \qquad a_0(u)v^n + a_1(u)v^{n-1} + \cdots + a_n(u) = 0.$$

Here $a_0(u) \neq 0$, and, because, when V is an indeterminate, V does not divide $F(u, V)$, $a_n(u) \neq 0$. Therefore, there is a rational number c such that both $a_0(c)$ and $a_n(c)$ are nonzero. Moreover, since $F(u, V)$ is irreducible, c can be chosen to make the discriminant of $F(c, V)$ nonzero as well, so that

$$a_0(c)V^n + a_1(c)V^{n-1} + \cdots + a_n(c)$$
$$= a_0(c)(V - \alpha_1)(V - \alpha_2) \cdots (V - \alpha_n)$$

where V is an indeterminate, where c is a rational number, where $a_0(c) \neq 0$, and where $\alpha_1, \alpha_2, \ldots, \alpha_n$ are distinct, nonzero elements of an algebraic extension of K_0. (If $F(u, V)$, regarded as a polynomial in V with coefficients in the field $K_0(u)$, had discriminant zero, then it would have a multiple root in an algebraic extension of $K_0(u)$, and F and its partial derivative with respect to V would have a nontrivial common factor over an algebraic extension of $K_0(u)$ and would therefore have a nontrivial common factor over $K_0(u)$ itself, contrary to the irreducibility of F and to unique factorization in $K_0[u, V]$.)

With such choices of u, v, $c \in K$, let j be the maximum of the degrees of the polynomials $a_i(u)$ ($i = 1, 2, \ldots, n$), let $w = u - c$, $x = 1/w$, $t = a_n(u)/v$, and $y = tx^j$. Multiplication of (1) by t^n and division by $a_n(u)$ gives

$$(2) \qquad a_0(u)a_n(u)^{n-1} + a_1(u)a_n(u)^{n-2}t + \cdots + t^n = 0.$$

Multiplication by x^{nj} then gives

$$x^{nj}a_0(u)a_n(u)^{n-1} + x^{(n-1)j}a_1(u)a_n(u)^{n-2}y + \cdots + y^n = 0,$$

which can be written

(3) $y^n + b_1(x)y^{n-1} + \cdots + b_n(x) = 0$

where

$$b_i(x) = x^{ij}a_{n-i}(u)a_n(u)^{i-1} = \left(x^j a_{n-i}(w+c)\right)\left(x^j a_n(w+c)\right)^{i-1}$$

is in $K_0[x]$ by the choice of j. By (3), y is integral over $\mathbf{Q}[x]$. By (2), $t = y/x^j$ is integral over $\mathbf{Q}[u] = \mathbf{Q}[u-c] = \mathbf{Q}[1/x]$. Moreover, y/x^j has n distinct, nonzero values where $x = \infty$. More precisely, $a_i(u)a_n(u)^{n-i-1} = a_i(w+c)a_n(w+c)^{n-i-1} = a_i(c)a_n(c)^{n-i-1} + w\psi_i(w)$, where ψ_i is a polynomial with coefficients in K_0, which gives, by (2), $0 = a_0(c)a_n(c)^{n-1} + a_1(c)a_n(c)^{n-2}t + \cdots + t^n + w\psi(w,t) = a_n(c)^{-1}t^n\left(a_0(c)a_n(c)^n t^{-n} + a_1(c)a_n(c)^{n-1}t^{-(n-1)} + \cdots + a_n(c)\right) + w\psi(w,t) = a_n(c)^{-1}t^n a_0(c)\left(a_n(c)t^{-1} - \alpha_1\right)\left(a_n(c)t^{-1} - \alpha_2\right)\cdots\left(a_n(c)t^{-1} - \alpha_n\right) + w\psi(w,t) = (-\alpha_1)(-\alpha_2)\cdots(-\alpha_n) \cdot a_n(c)^{-1}a_0(c)\left(t - \alpha_1^{-1}a_n(c)\right)\left(t - \alpha_2^{-1}a_n(c)\right)\cdots\left(t - \alpha_n^{-1}a_n(c)\right) + w\psi(w,t) = (t - \beta_1)(t - \beta_2)\cdots(t - \beta_n) + w\psi(w,t)$, where the $\beta_i = a_n(c)\alpha_i^{-1}$ are distinct, nonzero elements of the numerical extension K' of K obtained by adjoining $\alpha_1, \alpha_2, \ldots, \alpha_n$ to K, and where ψ is a polynomial in two indeterminates with coefficients in K_0. It follows directly from the definition of a place that $\{t - \beta_i, w\} = \{\frac{y}{x^j} - \beta_i, \frac{1}{x}\} = P_i$ is a place in K'. Since $\{1/x\}$ is a place in $K_0(x)$, $\deg_K\{1/x\} = [K : K_0(x)] \cdot 1 = n$ (see §3.21), and the places $P_1, P_2, \ldots, P_n$ account for all n places dividing $\{1/x\}$ in a numerical extension of K.

By Corollary 2 of §1.28, there is a $\Delta(x) \in \mathbf{Q}[x]$ such that every element of K integral over $\mathbf{Q}[x]$ can be written in the form $\phi(x,y)/\Delta(x)$, where ϕ is a polynomial with coefficients in $\mathbf{Q}$.

By (3), it follows that every element z of K integral over $\mathbf{Q}[x]$ can be written in the form $z = \sum_{i=1}^{n} \phi_i(x)y^{n-i}/\Delta(x)$, where the $\phi_i(x)$ are in $K_0[x]$. By division of polynomials, each $\phi_i(x)$ can be written in the form $\phi_i(x) = \theta_i(x)\Delta(x) + \xi_i(x)$, where $\deg \xi_i < \deg \Delta$, which puts z in the form $z = \sum \theta_i(x)y^{n-i} + \sum(\xi_i(x)/\Delta(x))y^{n-i}$.

Let $A = \{1/x\}^{-\nu}$ and let S be the set of all elements z of K divisible by A. Since $A_x = [1]_x$, every z in S is divisible by $[1]_x$ and is therefore of the form $z = \sum \theta_i(x)y^{n-i} + \sum(\xi_i(x)/\Delta(x))y^{n-i}$. Moreover, since the first sum on the right is integral over $\mathbf{Q}[x]$ (y is integral over $\mathbf{Q}[x]$), so is the second, which implies that the $n \deg \Delta$ coefficients of the ξ's satisfy a certain number of *linear* conditions. Therefore, the elements of the form $\sum(\xi_i/\Delta)y^{n-i}$ which are integral over $\mathbf{Q}[x]$ are a finite-dimensional vector space over K_0, that is, there is a finite set $\zeta_1, \zeta_2, \ldots, \zeta_k$ of elements of the form $\sum(\xi_i/\Delta)y^{n-i}$ integral over $\mathbf{Q}[x]$ such that every element of this form integral over $\mathbf{Q}[x]$ can be written in one and only one way as $a_1\zeta_1 + a_2\zeta_2 + \cdots + a_k\zeta_k$ with $a_i \in K_0$. Thus, every $z \in S$ can be written in one and only one way in the form $\sum \theta_i(x)y^{n-i} + \sum a_i\zeta_i$ with $\theta_i(x) \in K_0[x]$ and $a_i \in K_0$. (If $k = 0$, the second sum is absent.)

(The conditions that integrality of $\sum(\xi_i(x)/\Delta(x))y^{n-i}$ imposes on the coefficients of the ξ's can be described quite explicitly. Let L be a numerical extension of K in which $[\Delta(x)]$ is a product of powers of places. For each place P in the numerator of $[\Delta(x)]$, let $\sum \xi_i(x)y^{n-i}$ be expanded as a power series in a local parameter at P. The coefficient of each term is a linear combination of the coefficients of the ξ's, and the conditions are that a certain finite subset of these coefficients be zero.)

If $z \in S$, then the restriction to $1/x$ of $\{1/x\}^{\nu}[z]$ is integral, which is to say that z/x^{ν} is integral over $\mathbf{Q}[1/x]$. When z is written in the form $z = \sum \theta_i(x)y^{n-i} + \sum a_i\zeta_i$, the second sum divided by x^{ν} is integral over $\mathbf{Q}[1/x]$ provided ν is greater than some bound depending only on x. (Since ζ_i is integral over $\mathbf{Q}[x]$, the restriction of its denominator to x is $[1]_x$, so, by the proposition of §3.7, the denominator of ζ_i is concentrated at $\{1/x\}$, that is, $\{1/x\}^\nu\zeta_i$ is integral for ν large.) Therefore, provided ν is sufficiently large, $\sum\theta_i(x)y^{n-i}/x^\nu$ must be integral over $\mathbf{Q}[1/x]$. Since y/x^j is integral over $\mathbf{Q}[1/x]$, $\sum\theta_i(x)y^{n-i}/x^\nu = \sum\theta_i(x)x^{-\nu+j(n-i)}(y/x^j)^{n-i}$ is integral over $\mathbf{Q}[1/x]$ whenever $\deg\theta_i \leq \nu - j(n-i)$, because the coefficient of $(y/x^j)^{n-i}$ is then a polynomial in $1/x$. That this *sufficient* condition is also *necessary* can be seen as follows.

At each of the n distinct places P_σ where $x = \infty$ (i.e., $P_\sigma|\{1/x\}$), $y/x^j = \beta_\sigma + $ terms in x^{-1}, while $\theta_i(x) = x^{\deg\theta_i}(s_i + $ terms in $x^{-1})$, where s_i is the leading coefficient of $\theta_i(x)$. Therefore, $x^{-\nu}\theta_i(x)y^{n-i} = x^{-\nu+\deg\theta_i+j(n-i)} \cdot (s_i\beta_\sigma^{n-i} + $ terms in $x^{-1})$. Let μ be the maximum over i of $-\nu + \deg\theta_i + j(n-i)$. If $\mu > 0$ and if $x^{-\nu}\sum\theta_i(x)y^{n-i}$ is finite at P_σ, then the coefficients of x^μ must *cancel*, which is to say that $\sum s_i\beta_\sigma^{n-i} = 0$, where the sum is over all indices i for which $\mu = -\nu + \deg\theta_i + j(n-i)$. This is impossible because it states that n distinct constants $\beta_1, \beta_2, \ldots, \beta_n$ are roots of a nonzero (by the choice of μ) polynomial of degree less than n. Therefore, if $x^{-\nu}\sum\theta_i(x)y^{n-i}$ is integral over $\mathbf{Q}[1/x]$, μ must be ≤ 0, that is, $-\nu + \deg\theta_i + j(n-i) \leq 0$ for all i.

In summary, if $A|z$, then, provided ν is large, $z = \sum\theta_i(x)y^{n-i} + \sum a_i\zeta_i$, where $\deg\theta_i \leq \nu - j(n-i)$. Conversely, every element z of K of this form is integral over $\mathbf{Q}[x]$ and has the property that z/x^ν is integral over $\mathbf{Q}[1/x]$; thus, if $z \neq 0$, the restrictions of $\{1/x\}^\nu[z]$ to both x and $1/x$ are integral, which implies that $A|z$. Since the expression of z in this form contains $\sum_{i=1}^n(\nu - j(n-i) + 1) + k = n\nu - j\binom{n}{2} + n + k$ constants, this is $\dim_K(A)$. On the other hand, $\deg_K(A) = -\nu\deg_K\{1/x\} = -\nu n\deg_{\mathbf{Q}(x)}\{1/x\} = -\nu n$, so $\dim_K(A) + \deg_K(A) = -j\binom{n}{2} + n + k$. Let

$$g = j\binom{n}{2} - n - k + 1$$

so that $\dim_K(A) + \deg_K(A) + g = 1$. To prove the theorem, and to prove that g is the genus of K, it will suffice to prove that for any divisor B in K, $\dim_K(B) + \deg_K(B) \geq \dim_K(A) + \deg_K(A)$.

Let B be a divisor in K, and let C_x be the restriction to x of the denominator of B. There is a $g(x) \in \mathbf{Q}[x]$ such that $C_x|[g(x)]_x$ (§1.10). For such a $g(x)$, the restriction to x of $[g(x)]B$ is integral, which implies that $\{1/x\}^\nu[g(x)]B$ is integral for ν sufficiently large (§3.7), say $\{1/x\}^\nu[g(x)]B = D$ is integral. Since $B|z$ if and only if $[g(x)]B|g(x)z$, and since multiplication by $g(x)$ is invertible, $\dim_K(B) = \dim_K([g(x)]B)$. Since $\deg_K(B) = \deg_K([g(x)]B)$, it follows that the desired inequality $\dim_K(B) + \deg_K(B) \geq \dim_K(A) + \deg_K(A)$ will be proved if it is proved for $[g(x)]B = D/\{1/x\}^\nu = DA$. In short, with A as before (and with ν large), it will suffice to prove that $\dim_K(DA) + \deg_K(DA) \geq \dim_K(A) + \deg_K(A)$ for all integral divisors D in K. This can be done as follows.

Since there is a numerical extension L of K in which D is a product of places, and since $\dim_K(B) = \dim_L(B)$ and $\deg_K(B) = \deg_L(B)$ for all divisors B in K, one can assume without loss of generality that D is a product of places in K. Every z divisible by DA is divisible by A. Therefore, the vector space S' of elements of K divisible by DA is a *subspace* of the vector space S of elements divisible by A. In fact, $S' \subset S$ is described by $\deg_K D$ linear conditions. (At a place P which does not divide $\{1/x\}$, the expansion of an element $z \in S$ in powers of a local parameter at P has no terms of degree < 0. At a place P which does divide $\{1/x\}$, the expansion has no terms of degree $< -\nu$. If P is a place which divides D with multiplicity e, and if $z \in S'$, then the next e terms must also be absent—that is, there can be no terms of degree $< e$ when $P \nmid \{1/x\}$ and no terms of degree $< -\nu + e$ when $P|\{1/x\}$. The condition that these $\deg_K D$ expansion coefficients be zero is both necessary and sufficient for $z \in S'$.) Therefore, the dimension of S is at most $\deg_K D$ more than the dimension of S', that is, $\dim_K(A) \leq \dim_K(DA) + \deg_K D$. Addition of $\deg_K(A)$ on both sides of this inequality gives $\dim_K(A) + \deg_K(A) \leq \dim_K(DA) + \deg_K(DA)$, as was to be shown.

§3.25 Abel's Theorem.

Two divisors A and B in a function field K are *equivalent* if $A^{-1}B = [z]$ for some $z \in K$.

ABEL'S THEOREM. *In a function field K of genus g, every divisor of degree g or greater is equivalent to an integral divisor.*

PROOF: If $\deg_K A \geq g$, then $\dim_K(A^{-1}) = \dim_K(A^{-1}) + \deg_K(A^{-1}) + \deg_K(A) \geq \dim_K(A^{-1}) + \deg_K(A^{-1}) + g > 0$. Therefore, $A^{-1}|z$ for some nonzero $z \in K$, say $[z] = A^{-1}B$, and B is an integral divisor equivalent to A, as desired.

Since Abel's time, the name "Abel's Theorem" has been applied to a variety of statements, some of them quite remote from Abel. Abel's original statement [A] was roughly the following: Let C be an algebraic curve in the xy-plane, let $f(x,y)$ be a rational function on C, and let $\psi(P)$ denote the many-valued "function" $\psi(P) = \int_O^P f(x,y)\,dx$ on C, which depends on the lower limit of integration O and the choice of a path of integration from O to P, as well as on P. Abel stated that when $P_1, P_2, \ldots, P_\mu$ satisfy a certain set of g algebraic conditions (for some $g \geq 0$) the sum of μ integrals $\psi(P_1) + \psi(P_2) + \cdots + \psi(P_\mu)$ can be expressed as the integral of a *rational* function. He proved it by constructing a family of rational functions θ on C depending on a set of parameters $a, a', a'', \ldots$, such that the θ's all have the same μ poles, but their μ zeros depend on the parameters $a, a', a'', \ldots$. He then showed that the sum of $f(x_i(a), y_i(a))\,dx_i = f(x_i(a), y_i(a))x_i'(a)\,da$ over the μ zeros $(x_i(a), y_i(a))$ of the θ corresponding to a given value of the a's is a *rational* differential in the a's (because, in essence, this sum is a trace). Integration then gives

$$(1) \quad \sum_{i=1}^{\mu} \int_{Q_i}^{P_i} f(x, y)\, dx = \int \text{rational differential in the } a\text{'s}$$

where $Q_1, Q_2, \ldots, Q_\mu$ and $P_1, P_2, \ldots, P_\mu$ are the zeros on C of the θ's corresponding to two different sets of values of the parameters $a, a', a'', \ldots$. When the Q's are fixed, the left side of this equation is of the form $\sum \psi(P_i) + \text{const.}$ and the equation says that $\sum \psi(P_i)$ is an indefinite integral of a rational function of the a's, when the P's are the zeros of the θ corresponding to some set of values of $a, a', a'', \ldots$.

For any set of values of the parameters $a, a', a'', \ldots$, the zero set $P_1, P_2, \ldots, P_\mu$ of θ is equivalent, as a divisor, to the pole set, which is the same for all sets of values of the parameters. Therefore, the condition on the P's implies that the divisors $P_1 P_2 \cdots P_\mu$ are all equivalent. When the a's provide enough degrees of freedom, the *only* condition on the P's is that these divisors be equivalent. Thus, when $P_1 P_2 \cdots P_\mu$ is equivalent to a fixed divisor, $\sum \psi(P_i)$ can be expressed as an indefinite integral of a rational function of the a's.

From this point of view, the essence of Abel's original theorem is the statement that, given a set of places $Q_1, Q_2, \ldots, Q_\mu$ on an algebraic curve C, the nearby sets of places $P_1, P_2, \ldots, P_\mu$ such that the divisors $P_1 P_2 \cdots P_\mu$ and $Q_1 Q_2 \cdots Q_\mu$ are equivalent can be described by g algebraic conditions, for some g. The minimum value of g is the genus of C, and the desired conditions can be derived from "Abel's Theorem" above as follows. Given $P_1, P_2, \ldots, P_{\mu-g}$, there is, by "Abel's Theorem", an integral divisor of degree g, call it $R_1 R_2 \cdots R_g$, which is equivalent to $(Q_1 Q_2 \cdots Q_\mu)/(P_1 P_2 \cdots P_{\mu-g})$. The R's depend on $P_1, P_2, \ldots, P_{\mu-g}$, and the g conditions are $P_{\mu-g+i} = R_i$ for $i = 1, 2, \ldots, g$.

§3.26 The Genus as a Limit.

PROPOSITION. *Let A and B be divisors in a function field K, and let $\deg_K A$ be positive. Then $1 - \dim_K(A^{-\mu}B) - \deg_K(A^{-\mu}B)$ is the genus of K for all sufficiently large integers μ.*

PROOF: The case $A = \{1/x\}$, $B = [1]$ of this proposition, where x is a parameter of K of the special type considered in §3.24, was proved in §3.24. If A is any divisor with positive degree, and if ν is any integer, then $\deg_K(A^\mu B^{-1}\{1/x\}^{-\nu}) > g$ for μ sufficiently large. By Abel's theorem, $A^\mu B^{-1}\{1/x\}^{-\nu}$ is equivalent to an integral divisor, say $A^\mu B^{-1}\{1/x\}^{-\nu}[z] = C$, where C is integral and z is a nonzero element of K. When ν is large enough that $g = 1 - \dim_K(\{1/x\}^{-\nu}) - \deg_K(\{1/x\}^{-\nu})$, $1 - g = \dim_K(CB/A^\mu) + \deg_K(CB/A^\mu)$. The space of elements of K divisible by CB/A^μ is a subspace of the space of elements of K divisible by B/A^μ defined by $\deg_K(C)$ linear conditions. Therefore $\dim_K(CB/A^\mu) \geq \dim_K(B/A^\mu) - \deg_K(C)$, which combines with the equation above to give $1 - g \geq \dim_K(B/A^\mu) - \deg_K(C) + \deg_K(C) + \deg_K(B/A^\mu)$. Since $\dim_K(B/A^\mu) + \deg_K(B/A^\mu) \geq 1 - g$ by the definition of the genus, $1 - g = \dim_K(B/A^\mu) + \deg_K(B/A^\mu)$, as was to be shown.

COROLLARY. *The genus of a numerical extension of a function field K is equal to the genus of K.*

DEDUCTION: An integral divisor A in K is also an integral divisor in any numerical extension L of K. By the proposition, both the genus of K and the genus of L are equal to $1 - \dim_K(A^{-\mu}) - \deg_K(A^{-\mu})$ for all sufficiently large μ.

§3.27.

PROPOSITION. *Let K be a function field of genus g. There is an integral divisor A of degree g in a numerical extension K'*

of K with the property that the only integral divisor in K' or in a numerical extension of K' equivalent to A is A itself.

PROOF: As before, let K_0 be the field of constants of K, let x be a parameter of K, let ν be an integer large enough that $\dim_K \{1/x\}^{-\nu} = n\nu - g + 1$ where $n = \dim_K \{x\}$, and let S be the vector space of elements of K divisible by $\{1/x\}^{-\nu}$.

Let P_1, P_2, ..., P_j be a set of places in K which do not divide $\{1/x\}$ and which have the property that the mapping $S \to K_0^j$ obtained by evaluating elements of S at these j places is onto. (The empty set of places is to be regarded as satisfying this condition.) The main step in the proof is to show that if $j < n\nu - g + 1$ then, by passing to a numerical extension of K if necessary, one can find another place P_{j+1} such that evaluation at P_1, P_2, ..., P_{j+1} again gives an onto mapping $S \to K_0^{j+1}$. For this, note that $j < n\nu - g + 1$ means $\dim K_0^j < \dim S$, so there is a nonzero $z \in S$ whose values at P_1, P_2, ..., P_j are all zero. In a numerical extension K' of K, $\{1/x\}^\nu [z] = P_1 P_2 \cdots P_j Q_1 Q_2 \cdots Q_{n\nu-j}$, where the Q_i are places. Let c be a rational number different from the values of x at the places P_1, P_2, ..., P_j, Q_1, Q_2, ... $Q_{n\nu-j}$. (The value of x at Q_i may be ∞.) Let P_{j+1} be a place in a numerical extension K'' of K' which divides $\{x - c\}$. By the choice of c, P_{j+1} does not divide $\{1/x\}$ and z is nonzero at P_{j+1}. Let K be replaced by K''. Then $S \to K_0^j$ is still onto. Moreover, the map $S \to K_0^{j+1}$ is also onto, because its kernel is a proper subspace of the kernel of $S \to K_0^j$ (because z is not in it), so its image is of dimension greater than j.

Thus, by induction on j, K can be extended numerically to find a function field K and places P_1, P_2, ..., $P_{n\nu-g+1}$ in K such that evaluation at these places gives a map $S \to K_0^{n\nu-g+1}$ which is onto. Let z be a nonzero element of S whose values at P_1, P_2, ..., $P_{n\nu-g}$ are all zero, and let $\{1/x\}^\nu [z] = P_1 P_2 \cdots P_{n\nu-g} Q_1 Q_2 \cdots Q_g$. Let B be an integral divisor in a numerical extension of K which is equivalent to $Q_1 Q_2 \cdots Q_g$.

Then $P_1 P_2 \cdots P_{n\nu-g} B\{1/x\}^{-\nu}$ is equivalent to $[z]$, which is equivalent to $[1]$, so there is a w in a numerical extension of K such that $\{1/x\}^{\nu}[w] = P_1 P_2 \cdots P_{n\nu-g} B$. Thus, when K is extended numerically, w is in S and is in the kernel of the map $S \to K_0^{n\nu-g}$ given by evaluation at $P_1, P_2, \ldots, P_{n\nu-g}$. This kernel is one-dimensional, so w is z times an element of K_0. Thus, $[w] = [z]$, which implies $B = Q_1 Q_2 \cdots Q_g$. Therefore $A = Q_1 Q_2 \cdots Q_g$ is a divisor with the required properties.

COROLLARY. *There is a divisor of degree $g - 1$ in a numerical extension of K which is not equivalent to an integral divisor.*

(It follows, by Abel's theorem, that g can be characterized as the largest integer such that some divisor of degree $g - 1$ in some numerical extension of K is not equivalent to an integral divisor.)

DEDUCTION: Let A be as in the proposition, and let P be a place in a numerical extension of K which does not divide A. Then A/P has the required property.

§3.28 The Divisor Class Group.

Equivalence of divisors (§3.25) is obviously consistent with multiplication of divisors, and the resulting multiplication of equivalence classes obviously gives to them the structure of a commutative group. This is the *divisor class group* of K.

Equivalent divisors have equal degrees, so equivalence classes of divisors have degrees. The assignment to divisor classes of their degrees is a homomorphism from the divisor class group to **Z**.

Let O be a divisor in K of degree g. Every divisor of degree zero in K is, by Abel's theorem, of the form A/O where A is an integral divisor of degree g. Therefore, the mapping which sends an integral divisor A of degree g to the class of A/O is a map of such divisors onto the subgroup of the divisor class group consisting of divisor classes of degree zero. Thus, elements of the divisor class group of degree zero can be identified, given O, with equivalence classes of integral divisors of degree g. When this identification is made, the group law $(A/O)(B/O) \sim C/O$ takes the form

(1) $$AB \sim CO$$

that is, the product of the elements of the divisor class group represented by A and B (relative to O) is represented by any integral divisor C for which (1) is satisfied. Abel's theorem implies that for integral divisors A and B of degree g, and for a divisor O of degree g, there is an integral divisor C for which (1) holds.

§3.29 Examples.

If $K = \mathbf{Q}(x)$ and $A = \{1/x\}$, then $z \in K$ is divisible by $A^{-\nu}$ if and only if z is a polynomial of degree ν, at most, in x. The genus of this K is therefore equal to $1 - \dim_K(\{1/x\}^{-\nu}) - \deg_K(\{1/x\}^{-\nu}) = 1 - (\nu + 1) - \nu = 0$, as is the genus of any numerical extension of $\mathbf{Q}(x)$.

Let K be a function field of genus 0. Let P and Q be distinct places in a numerical extension L of K. By Abel's theorem, P/Q is equivalent to an integral divisor, say $[z] = P/QR$, where R is integral. Then $\{z\} = P$, which gives

$$1 = \deg_L\{z\} = [L \colon \mathbf{Q}(z)][L_0 \colon \mathbf{Q}]^{-1} \deg_{\mathbf{Q}(z)}\{z\}$$
$$= [L \colon L_0(z)][L_0(z) \colon \mathbf{Q}(z)][L_0 \colon \mathbf{Q}]^{-1} = [L \colon L_0(z)]$$

(§3.21), where L_0 is the field of constants of L. Therefore, $L = L_0(z)$ is a numerical extension of $\mathbf{Q}(z)$. Thus, *a function field has genus 0 if and only if it has a numerical extension isomorphic to a numerical extension of* $\mathbf{Q}(x)$.

Let K be a function field of genus 1, and let P, Q, and R be distinct places in a numerical extension L of K. By Abel's theorem, PQ/R is equivalent to an integral divisor, say $[z] = PQ/RS$, where S is integral. Then $\{z\} = PQ$, and, by the computation above, $2 = \deg_L\{z\} = [L:L_0(z)]$, which shows that L is a quadratic extension of a numerical extension $L_0(z)$ of $\mathbf{Q}(z)$. Thus, a generator y of L over $L_0(z)$ satisfies an equation of the form $a_0(z)y^2 + a_1(z)y + a_2(z) = 0$. Then $w = 2a_0(z)y + a_1(z)$ is also a generator of L over $L_0(z)$, and $w^2 = a_1(z)^2 - 4a_0(z)a_2(z)$. Therefore, a numerical extension of K is equal to the field obtained by adjoining the square root of a polynomial to a numerical extension of $\mathbf{Q}(z)$.

Let L_0 be an algebraic number field, let $b(x) \in L_0[x]$, and let L be the field obtained by adjoining a root y of $y^2 - b(x)$ to $L_0(x)$. If $b(x)$ has odd degree $2j - 1$, and if c is a rational number such that $b(c) \neq 0$, then, with $x = c + u^{-1}$, $(u^j y)^2 = u^{2j}b(c + u^{-1})$ is a polynomial in u of degree $2j$. Thus, L can be obtained by adjoining to $L_0(x) = L_0(u)$ a square root of a polynomial of degree $2j$, and there is no loss of generality in assuming that the original $b(x)$ is of even degree $2j$. By passing to a numerical extension of L, if necessary, one can also assume that $b(x)$ is a product of *monic*, *linear* factors $b(x) = (x - e_1)(x - e_2) \cdots (x - e_{2j})$. Finally, the factors $x - e_i$ can be assumed to be *distinct*, because adjoining a square root of $(x - e)^2 b(x)$ is the same as adjoining a square root of $b(x)$. In summary, every function field K of genus 1 has a numerical extension of the form $L_0(x, \sqrt{b(x)})$, where $b(x) = (x - e_1)(x - e_2) \cdots (x - e_{2j})$ is a product of distinct factors.

The genus of a function field of the form $L_0(x, \sqrt{b(x)})$ just described is $j - 1$, as can be proved as follows. Because $(y/x^j)^2 = 1 + $ terms in x^{-1}, the values of y/x^j at $x = \infty$ are ± 1, and $g = j\binom{2}{2} - 2 - k + 1 = j - 1 - k$ by the formula of §3.24, where k is the dimension over L_0 of the vector space of elements of L of the form $(\phi(x) + y\psi(x))/\Delta(x)$ which are integral over $\mathbf{Q}[x]$ ($\deg \phi < \deg \Delta$, $\deg \psi < \deg \Delta$). If $z = (\phi(x) + y\psi(x))/\Delta(x)$ is integral over $\mathbf{Q}[x]$, then $\phi(x) + y\psi(x) - z\Delta(x) = 0$, where $1, y$ are a basis of L over the field of quotients $L_0(x)$ of the natural ring $L_0[x]$, where $1, y, z$ are integral over $L_0[x]$, and where $\phi(x), \psi(x), \Delta(x) \in L_0[x]$ are relatively prime. By Corollary 1 of §1.30, $\Delta(x)^2 | \mathrm{disc}_L(1, y)$. By the proposition of §1.30, $\mathrm{disc}_L(1, y)$ is the norm of $[F'(y)]$ where $F(X) = N_{L/L_0(x)}(X - y) = X^2 - b(x)$, that is, $\mathrm{disc}_L(1, y) = N_{L/L_0(x)}[2y] = [-4b(x)]$. Since $b(x)$ has no square factors, it follows that $\Delta(x)$ is a unit of $L_0[x]$, and therefore that $\phi(x)$ and $\psi(x)$ are 0. Therefore, $k = 0$ and $g = j - 1$, as claimed.

In conclusion, then, every function field of genus 1 has a numerical extension of the form $L_0(x, \sqrt{b(x)})$, where L_0 is an algebraic number field and $b(x)$ is a product of 4 distinct factors $(x - e_1)(x - e_2)(x - e_3)(x - e_4)$ with $e_i \in L_0$. Conversely, any function field with a numerical extension of this form has genus 1. Such a function field is called *elliptic*.

In the case of an elliptic function field K, since $g = 1$, the construction of §3.28 gives a group structure to the set of *places* in K, once a place O is chosen to play the role of the identity. The usual geometrical description of this group of places is the following. A simple change of variables can be used to put the defining equation of K in the form $y^2 = a_1 x^3 + a_2 x^2 + a_3 x + a_4$ where the a_i are constants of K and $a_1 \neq 0$. (Given the defining equation in the form $y^2 = (x - e_1)(x - e_2)(x - e_3)(x - e_4)$, set $x' = (x - e_4)^{-1}$ and $y' = y(x - e_4)^{-2}$.) With $u = 1/x$ and $v = y/x^2$, one then finds $v^2 = a_1 u + a_2 u^2 + a_3 u^3 + a_4 u^4$, from which it follows that $\{u, v\}$ is a place in K at which v is a local parameter and at which u has a double zero. Therefore, with $O = \{u, v\}$, x has a double pole at O and $y = vx^2$ has a triple pole there. In fact, the denominator of $[x]$ is O^2, because $\{x\}$ is a place in $K_0(x) \subset K$ and this extension has degree $[K : K_0(x)] = 2$ (the extension is generated by the element y of degree 2 over $K_0(x)$), so $\deg_K\{1/x\} = \deg_K\{x\} = 2$ (§3.21). Similarly, $\{1/y\} = O^3$. Thus, for any b_1, b_2, b_3 in the field of constants K_0 of K, the denominator of the divisor of $\phi = b_1 + b_2 x + b_3 y$ is O^3 provided $b_3 \neq 0$. The denominator of the divisor of $\phi = b_1 + x$ is O^2. Given two places P, Q in K (not necessarily distinct, but neither equal to O) there exist b_1, b_2,

$b_3 \in K_0$ such that $\phi = b_1 + b_2 x + b_3 y$ is not constant and is zero at P and Q (2 homogeneous linear equations in 3 unknowns of which 1, 0, 0 is not a solution). Thus $[\phi] = PQR/O^3$, where R is a place in K. Geometrically, R is the third point in which the line $\phi = 0$ in the xy-plane through P and Q intersects the given curve. (If $b_3 = 0$, then $R = O$.) Let c be the value of x at R (unless $R = O$) and let the equation $[x - c] = RS/O^2$ define a place S on the curve. Geometrically, S is the other point in which the vertical line in the xy-plane through R intersects the curve. Then $PQR/O^3 \sim [1]$ and $RS/O^2 \sim [1]$, which gives $PQR \sim O^3 \sim RSO$. Therefore, $PQ \sim OS$, which means that S is the product of P and Q in the group in question. (If $R = O$, then $PQO \sim O^3$, $PQ \sim O^2$, so O is the product of P and Q. If either P or Q is O, their product is the other one.)

A function field of the form $L_0(x, \sqrt{b(x)})$ in which $b(x)$ is a product of more than 4 distinct linear factors in $L_0[x]$ is called *hyperelliptic*. As the above argument shows, there are hyperelliptic function fields with any genus greater than 1.

A simple example of a function field of higher genus which is not hyperelliptic is the field K of rational functions on the so-called *Klein curve* defined by the equation $y^3 + x^3 y + x = 0$. Clearly $P = \{x, y\}$ is a place in K at which y is a local parameter and at which x has a zero of order 3. In fact, since $\deg_K\{x\} = 3 \deg_{Q(x)}\{x\} = 3$, $\{x\} = P^3$. Writing the equation of the Klein curve in the homogeneous form $Y^3 Z + X^3 Y + Z^3 X = 0$ with $y = Y/Z$ and $x = X/Z$ brings out a symmetry which shows that $Q = \{Z/Y, X/Y\} = \{1/y, x/y\}$ is a place in K where x/y is a local parameter and $1/y$ has a zero of order 3, while $R = \{Y/X, Z/X\} = \{y/x, 1/x\}$ is a place in K where $1/x$ is a local parameter and y/x has a zero of order 3. Thus $1/x$ has a simple zero at R and, since $1/x = (1/y)/(x/y)$, it has a double zero at Q. Therefore, $\{1/x\} = Q^2 R$. It will be shown that the genus of K is 3 by applying the Proposition of §3.26 in the case of $A = Q^2 R = \{1/x\}$ and $B = [1]$.

Let S be the subspace of elements of K divisible by $A^{-\mu}$ for some large μ. An element z of K is in S if and only if z is integral over $\mathbf{Q}[x]$ and has poles of order 2μ at most at Q and of order μ at most at R. Because 1, y, y^2 are a basis of K over $\mathbf{Q}(x)$, z satisfies a relation of the form $\phi(x)z = \xi_1(x)+\xi_2(x)y+\xi_3(x)y^2$, where the coefficients $\phi(x)$ and $\xi_i(x)$ are in $\mathbf{Q}[x]$ and are relatively prime. One can assume, moreover, that $\phi(x)$ is monic. By Corollary 1 of §1.30, $\phi(x)^2$ divides $\mathrm{disc}(1,\ y,\ y^2)$. By the proposition of §1.30, this discriminant is $N_K(3y^2+x^3)$, which is

$$\begin{vmatrix} x^3 & 0 & 3 \\ -3x & -2x^3 & 0 \\ 0 & -3x & -2x^3 \end{vmatrix} = 4x^9 + 27x^2.$$

Since the only monic square factor of this discriminant is x^2, $\phi(x)$ is 1 or x. If it were x, then $\xi_1(x) + \xi_2(x)y + \xi_3(x)y^2$ would be zero at P which is only possible, because y has a simple zero at P and x has a zero of order 3 there, when the ξ_i are all divisible by x, contrary to the assumption that $\phi(x)$ and the $\xi_i(x)$ are relatively prime. Therefore $z = \xi_1(x) + \xi_2(x)y+\xi_3(x)y^2$. Let a, b, c be the degrees of ξ_1, ξ_2, ξ_3. At the place Q, x has a pole of order 2 and $1/y$ has a zero of order 3, that is, y has a pole of order 3. The individual terms of $z = \xi_1(x) + \xi_2(x)y + \xi_3(x)y^2$ therefore have poles of order $2a$, $2b + 3$, $2c + 6$ respectively at Q, while z has a pole of order 2μ, at most, at Q. At the place R, x has a pole of order 1 and $y = (y/x)/(1/x)$ has a zero of order 2, so the individual terms of z have poles of order a, $b-2$, $c-4$, respectively, while z has a pole of order μ, at most. If a were greater than μ, $\xi_1(x)$ would have a pole of order $2a > 2\mu$ at Q. Therefore, $\xi_2(x)y+\xi_3(x)y^2$ would have to have a pole of order exactly $2a$ at Q; since $2b+3$ is odd, it would follow that $2c + 6 = 2a$ and $2c + 6 > 2b + 3$. But then z would have a pole of order $a > \mu$ at R, because $a > b$ and $a > c$. Since this is not the case, $a \le \mu$. Then, since $\xi_2(x)y+\xi_3(x)y^2$ has a pole of order 2μ at most at Q, and since $2b + 3$ is odd and $2c + 6$ is even, $2b + 3 \le 2\mu$ and $2c + 6 \le 2\mu$.

Thus, $a \leq \mu$, $b \leq \mu - 2$, $c \leq \mu - 3$. Conversely, for such a, b, c, the sum $z = \xi_1(x) + \xi_2(x)y + \xi_3(x)y^2$ is in S because each term is in S. Thus $g = 1 - \dim_K(A^{-\mu}) - \deg_K(A^{-\mu}) = 1 - ((\mu + 1) + (\mu - 1) + (\mu - 2)) + 3\mu = 3$.

Appendix: Differentials

§A.1.

Let P_1, P_2, ..., P_k be places in a function field K. The poles of an element z of K will be said to be concentrated at $P_1 P_2 \cdots P_k$ if $\{1/z\}$ is concentrated at $P_1 P_2 \cdots P_k$ (see §3.6). Let u_i be a local parameter at P_i ($i = 1, 2, ..., k$). The condition "$z = \eta_i + a_1^{(i)} u_i^{-1} + a_2^{(i)} u_i^{-2} + \cdots$ where η_i is finite at P_i" determines a polynomial $a_1^{(i)} X + a_2^{(i)} X^2 + \cdots = \phi_i(X)$ in X with coefficients in the field of constants K_0 of K and with constant term zero. If the poles of z are concentrated at $P_1 P_2 \cdots P_k$, the polynomials $\phi_1(X)$, $\phi_2(X)$, ..., $\phi_k(X)$ are called the *principal parts* of z (relative to the parameters u_i). The problem dealt with in this appendix is: Given a set of polynomials $\phi_1(X)$, $\phi_2(X)$, ..., $\phi_k(X)$ (with coefficients in the field of constants K_0 of K and with constant terms zero) when are they the principal parts (relative to the u_i) of an element of K with poles concentrated at $P_1 P_2 \cdots P_k$? Using *differentials* one can find conditions on k-tuples (ϕ_1, ϕ_2, ..., ϕ_k) which are necessary and sufficient for them to be principal parts. The Riemann-Roch theorem is a simple corollary of this description of the principal parts.

§A.2 Definitions and First Propositions.

Let K be a function field. *Differentials* in K are represented by expressions of the form $x_1 dx_2 + x_3 dx_4 + \cdots + x_{2k-1} dx_{2k}$, where k is a positive integer and $x_1, x_2, ..., x_{2k}$ are elements of K. Such an expression *represents the zero differential*, denoted $x_1 dx_2 + x_3 dx_4 + \cdots + x_{2k-1} dx_{2k} = 0$, if, at every place P of every numerical extension of K, either (1) one of $x_1, x_2, ..., x_{2k}$ is not finite at P, or (2) they are all finite at P, and, for a local parameter t at P, when constants a_i, b_i are defined by $x_i = a_i + b_i t + o(t)$, the equation $a_1 b_2 + a_3 b_4 + \cdots + a_{2k-1} b_{2k} = 0$ holds. (The equation $x_i = a_i + b_i t + o(t)$ means, of course, that $(x_i - a_i - b_i t)/t$ is zero at P.) If $a_1 b_2 + a_3 b_4 + \cdots + a_{2k-1} b_{2k} = 0$

holds for one local parameter t at P, then it holds for all, because, if s is another, $s = \gamma t + o(t)$, $t = \gamma^{-1} s + o(s)$ where γ is a constant of K, $\gamma \neq 0$, and $x_i = a_i + b_i(\gamma^{-1} s + o(s)) + o(s) = a_i + \gamma^{-1} b_i s + o(s)$, so that the desired relation is $\sum a_{2i-1} \gamma^{-1} b_{2i} = \gamma^{-1} \sum a_{2i-1} b_{2i} = 0$, which holds whenever $\sum a_{2i-1} b_{2i} = 0$. Two expressions $\sum x_{2i-1} dx_{2i}$ and $\sum y_{2i-1} dy_{2i}$ *represent the same differential*, denoted $\sum x_{2i-1} dx_{2i} = \sum y_{2i-1} dy_{2i}$, if $\sum x_{2i-1} dx_{2i} + \sum (-y_{2i-1}) dy_{2i}$ represents the zero differential.

PROPOSITIONS. (1) $\sum x_{2i-1} dx_{2i} = \sum y_{2i-1} dy_{2i}$ *is an equivalence relation consistent with addition and with multiplication by elements of K.*

(2) *If $x \in K$ is constant, then $dx = 0$.*

(3) $d(x + y) = dx + dy$.

(4) $d(xy) = x\, dy + y\, dx$.

(5) *Let a parameter x of K be given. There is an integral divisor C in K such that, for every place P in every numerical extension of K, either x is not finite at P or $x - a$ is a local parameter at P for some constant a of K or $P|C$.*

(6) *If $dx = 0$, then x is a constant of K.*

(7) *Given $x_1, x_2, \ldots, x_{2k}$ and y in K, where y is not constant, $\sum x_{2i-1} dx_{2i} = z\, dy$ for a unique z in K.*

(8) *If K and L are function fields with $K \subset L$, and if $\sum x_{2i-1} dx_{2i} = \sum y_{2i-1} dy_{2i}$ holds in K, then $\sum x_{2i-1} dx_{2i} = \sum y_{2i-1} dy_{2i}$ holds in L.*

Note that Proposition (7) gives an algorithm for determining whether a given differential is zero; one need only write it in the form $z\, dy$, where y is not constant, and note that, by (1), (2), and (6), $z\, dy = 0$ if and only if $z = 0$.

PROOFS: (1) These properties are all easy to deduce from the definition. (2) If x is constant then x is finite at all places P in all numerical extensions of K and its expansion in terms of any local parameter t at P is $x = x + 0 \cdot t + 0$, so $\sum a_{2i-1} b_{2i}$ in the case of $1 \cdot dx$ is $1 \cdot 0 = 0$ for all P. (3) and (4) At any place P in any numerical extension of K at which x and y are both finite, $x = a + bt + w$ and $y = a' + b't + w'$, where t is a local parameter at P, $\operatorname{ord}_P(w) > 1$, $\operatorname{ord}_P(w') > 1$, and a, a', b, b' are constants. Then $x + y = (a + a') + (b + b')t + (w + w')$ and $xy = aa' + (ab' + a'b)t + (aw' + a'w)$, from which $d(x + y) - dx - dy = 0$ and $d(xy) - x\,dy - y\,dx = 0$ follow. (5) Since x is a parameter of K, K is generated over $\mathbf{Q}(x)$ by an element $y \in K$ which satisfies an equation of the form $F(x, y) = 0$, where $F(X, Y) = Y^n + c_1(X)Y^{n-1} + \cdots + c_n(X)$ is a polynomial with coefficients which are constants of K (§3.2, §1.25). Let $F_Y(X, Y) = nY^{n-1} + (n-1)c_1(X)Y^{n-2} + \cdots + c_{n-1}(X)$. Since $F_Y(x, y) \neq 0$ (F can be assumed to have minimal degree in Y), $\{F_Y(x, y)\}$ is an integral divisor. Let $C = \{F_Y(x, y)\}$. Let P be a place in a numerical extension K' of K at which x is finite. Then, since y is integral over $\mathbf{Q}[x]$, y is finite at P. Then $P|\{x - a, y - a'\}$ where a and a' are the values of x and y, respectively, at P. Expansion of $0 = F(x, y) = F(a + (x - a), a' + (y - a'))$ gives a polynomial relation $0 = \alpha + \beta(x - a) + \gamma(y - a') + \cdots$ satisfied by $x - a$ and $y - a'$, where $\alpha = F(a, a')$ is the value of $F(x, y) = 0$ at P—that is, $\alpha = 0$—and $\gamma = F_Y(a, a')$ is the value of $F_Y(x, y)$ at P. If $P \nmid C$ then $\gamma \neq 0$ and $\{x - a, y - a'\}$ is, directly from the definition, a place in K'. Since $P|\{x - a, y - a'\}$, it follows that $P = \{x - a, y - a'\}$. If P divided $x - a$ twice, then all terms of $0 = \beta(x - a) + \gamma(y - a') + \cdots$ would be divisible by P^2 with the possible exception of $\gamma(y - a')$, which would imply $P^2|\gamma(y - a')$ and therefore $P^2|\{x - a, y - a'\} = P$, $P = [1]$, which is not the case. Thus, $\operatorname{ord}_P(x - a) = 1$, and $x - a$ is a local parameter at P whenever $P \nmid C$ and $P \nmid \{1/x\}$, as was to be shown. (6) Assume x is not constant and let C be as in (5). Let z be a parameter of K which supports $\{1/x\}C$ (§3.8) and let P be a place in a numerical extension K' of K which divides $\{1/z\}$ (§3.18). Then P is relatively prime to $\{1/x\}$ and C, so x is finite at P and $P \nmid C$. Therefore, by (5), $x - a$ is a local parameter at P for some constant a of K' (the value of x at P). The equation $x = a + (x - a)$ then shows that $dx = 1 \cdot dx$ does not represent the zero differential. Therefore, x not constant implies $dx \neq 0$, as was to be shown. (7) Consider first the case $k = 1$, $x_1 = 1$. Since K is an algebraic extension of $\mathbf{Q}(y)$, x_2 and y satisfy a relation of the form $G(x_2, y) = 0$, where G is a polynomial with coefficients in $\mathbf{Z}$. Since 0 is constant, $d(G(x_2, y)) = 0$ by (2). By (2), (3), and (4), $d(G(x_2, y))$ represents the same differential as $G_X(x_2, y)dx_2 + G_Y(x_2, y)dy$, where G_X and G_Y are polynomials, and the degree of $G_X(X, Y)$ in X is less than that of $G(X, Y)$. Because the degree of G in X can be assumed to be minimal, $G_X(x_2, y)$ can be assumed to be nonzero. Thus, $dx_2 = z\, dy$, where $z = -G_Y(x_2, y)/G_X(x_2, y)$. For $k \geq 1$, $dx_{2i} = z_{2i}dy$, where $z_2, z_4, \ldots, z_{2k} \in K$, and $x_1 dx_2 + \cdots + x_{2k-1}dx_{2k} = z\, dy$ where $z = x_1 z_2 + \cdots + x_{2k-1}z_{2k}$. That z is uniquely determined follows from (6). (8) It has been shown that $z\, dy = 0$ if and only if $z = 0$ or $y = $ constant. Therefore, $z\, dy = 0$ in K if and only if $z\, dy = 0$ in L.

§A.3 Orders and Residues of Differentials.

Let $\sum x_{2i-1}dx_{2i}$ represent a differential in a function field K, let P be a place in K, and let t be a local parameter at P. By Proposition (6) above, $\sum x_{2i-1}dx_{2i} = u\, dt$ for some $u \in K$. The *order* of the differential $\sum x_{2i-1}dx_{2i}$ at P relative to t is defined to be $\mathrm{ord}_P(u)$.

PROPOSITION (9). *The order of a differential at P relative to t is the same for all local parameters t at P.*

DEFINITION. The order of a differential $\sum x_{2i-1}dx_{2i}$ at P, denoted $\mathrm{ord}_P(\sum x_{2i-1}dx_{2i})$, is its order at P relative to a local parameter at P.

PROOF: Let $P = \{x, y\}$ be a presentation of P as a place (§3.13). Assume without loss of generality that x is a local parameter at P, and let $z \in K$ be such that $z\, dx$ represents the given differential. If t is any local parameter at P, then $\mathrm{ord}_P(t/x) = \mathrm{ord}_P t - \mathrm{ord}_P x = 1 - 1 = 0$, so $tx^{-1} = a + \delta(x, y)(c + \varepsilon(x, y))^{-1}$, where a and c are nonzero constants of

K and δ and ε are polynomials with no constant term. Multiplication of this relation by $x(c + \varepsilon(x, y))$ gives $t(c + \varepsilon(x, y)) = xa(c + \varepsilon(x, y)) + x\delta(x, y)$. Thus $c\,dt + \varepsilon(x, y)dt + f_1(x, y, t)dx + f_2(x, y, t)dy = ac\,dx + g_1(x, y)dx + g_2(x, y)dy$, where f_1, f_2, g_1, g_2 are polynomials with no constant term. On the other hand, differentiation of the relation $F(x, y) = 0$ which presents $\{x, y\}$ as a place gives $F_X\,dx + F_Y\,dy = 0$, where F_X and F_Y are polynomials in x and y and where F_Y has nonzero constant term. Thus, $dy = q\,dx$ where $q = -F_X/F_Y$ is finite at P. Since $(c + \varepsilon)dt = (ac - f_1 - qf_2 + g_1 + qg_2)dx$, it follows that $dt = q_1\,dx$ where $q_1 = (ac - f_1 - qf_2 + g_1 + qg_2)/(c + \varepsilon)$ is finite, with value $ac/c = a \neq 0$, at P. Thus, the given differential $\sum x_{2i-1}dx_{2i}$ is represented by $zq_1^{-1}dt$. Since $\mathrm{ord}_P(zq_1^{-1}) = \mathrm{ord}_P z - \mathrm{ord}_P q_1 = \mathrm{ord}_P z$, the proposition follows.

Note that if $u\,dy$ is a differential in K, if $z \in K$, and if P is a place in K, then $\mathrm{ord}_P(zu\,dy) = \mathrm{ord}_P(z) + \mathrm{ord}_P(u\,dy)$.

PROPOSITION (10). *Let P be a place in a function field K, and let $z \in K$ have order $\mu > 0$ at P. Then dz has order $\mu - 1$ at P.*

PROOF: If x is a local parameter at P, then $z = cx^\mu + w$, where c is a nonzero constant of K and $\mathrm{ord}_P w = \nu > \mu > 0$. Let $w' = w/x^{\nu-1}$. Then $z = cx^\mu + x^{\nu-1}w'$, $dz = c\mu x^{\mu-1}dx + (\nu - 1)x^{\nu-2}w'dx + x^{\nu-1}dw'$. Since w' and x are local parameters at P, the last two differentials in the sum on the right have order $\nu - 1$ at P. The first has order $\mu - 1 < \nu - 1$, so dz has order $\mu - 1$, as was to be shown.

Let $u\,dy$ represent a differential in a function field K, and let t be a local parameter at P. The *residue* of $u\,dy$ at P relative to t is defined to be the coefficient of t^{-1} in the expansion of v in powers of t, where v is defined by $u\,dy = v\,dt$.

PROPOSITION (11). *The residue of a differential at P relative to t is the same for all local parameters t at P.*

DEFINITION. The residue of a differential $u\,dy$ at P, denoted

$\mathrm{res}_P(u\,dy)$, is its residue at P relative to a local parameter at P.

PROOF: Note first that if $\mathrm{ord}_P(u\,dy) \geq 0$ then the residue of $u\,dy$ at P relative to t is 0 for any local parameter t at P. Assume, therefore, that $\mathrm{ord}_P(u\,dy) = \nu < 0$, say $u\,dy = (a_\nu t^\nu + a_{\nu-1}t^{\nu-1} + \cdots + a_{-1}t^{-1} + w)dt$, where t is a local parameter at P, the a_i are constants of K with $a_\nu \neq 0$, and $\mathrm{ord}_P w \geq 0$. If t' is another local parameter at P, and if $\mathrm{res}'_P(v\,dz)$ denotes the residue of $v\,dz$ at P relative to t', then $\mathrm{res}'_P(u\,dy) = a_\nu\,\mathrm{res}'_P(t^\nu dt) + \cdots + a_{-1}\,\mathrm{res}'_P(t^{-1}dt) + \mathrm{res}'_P(w\,dt)$. The last term on the right is zero because $\mathrm{ord}_P(w\,dt) \geq 0$. That $\mathrm{res}'_P(t^j dt) = 0$ for $j < -1$ can be proved as follows. Let $t^{j+1}/(j+1)$ be expanded in powers of t', say $t^{j+1}/(j+1) = c_k(t')^k + c_{k+1}(t')^{k+1} + \cdots + c_N(t')^N + s$, where the c's are constants of K and $\mathrm{ord}_P s > N$. Then $t^j dt = d(t^{j+1}/(j+1)) = (kc_k(t')^{k-1} + \cdots + Nc_N(t')^{N-1})dt' + ds$. Since $kc_k(t')^{k-1} + \cdots + Nc_N(t')^{N-1}$ has no term in $(t')^{-1}$, and since $\mathrm{res}'_P(ds) = 0$ when $N \geq 0$, $\mathrm{res}'_P(t^j dt) = 0$. Finally, $t/t' = a + v$, where a is a nonzero constant of K and $\mathrm{ord}_P v > 0$. Then $t = (a + v)t'$ and $dt/t = d(a + v)/(a + v) + dt'/t'$. Since $\mathrm{ord}_P(a + v) = 0$, $\mathrm{ord}_P(d(a+v)/(a+v)) = \mathrm{ord}_P(d(a+v)) = \mathrm{ord}_P(dv) = \mathrm{ord}_P v - 1 > -1$. Therefore, $\mathrm{res}'_P(dt/t) = \mathrm{res}'_P(dt'/t') = 1$, and $\mathrm{res}'_P(u\,dy) = a_{-1} = \mathrm{res}_P(u\,dy)$, as was to be shown.

§A.4 A Fundamental Theorem.

The sum of the residues of a differential in a function field is zero. The following proposition makes possible a precise formulation of this basic fact.

PROPOSITION. *Let $u\,dx$ represent a differential in a function field K. There is an integral divisor B in K such that every place P in every numerical extension of K satisfies either $P|B$ or $\mathrm{ord}_P(u\,dx) = 0$.*

THEOREM. *With K, $u\,dx$, and B as in the proposition, let K' be a numerical extension of K in which B is a product* *of places. Then*

$$\sum_{P|B} \mathrm{res}_P(u\,dx) = 0$$

where the sum is over all places of K' which divide B.

PROOF OF THE PROPOSITION: Let C be an integral divisor as in (5) of §A.2, and let $B = \{u\}\{1/u\}\{1/x\}C$. If P is a place in a numerical extension of K which does not divide B then u and x are both finite at P and u is nonzero at P. Moreover, $x - a$ is a local parameter at P, where a is the value of x at P, so $dx = d(x - a)$ has order 0 at P. Therefore, $\mathrm{ord}_P(u\,dx) = \mathrm{ord}_P(u) + \mathrm{ord}_P(dx) = 0$, as was to be shown.

The proof of the theorem will use three lemmas.

LEMMA 1. *Let u and x be elements of a function field K such that $u\,dx$ is a differential whose poles are concentrated at $\{1/x\}$. Then the trace of u relative to the extension $K \supset K_0(x)$ is in $K_0[x]$. (Here K_0 is the field of constants of K, $K_0[x] \subset K$ is the ring of polynomials in x with coefficients in K_0, and $K_0(x) \subset K$ is the field of quotients of $K_0[x]$.)*

PROOF: Let the trace of u relative to $K \supset K_0(x)$ be $\phi(x)/\psi(x)$, where $\phi(x)$, $\psi(x) \in K_0[x]$ are relatively prime. In a numerical extension K' of K, $\{\psi(x)\}$ is a product of places. If the poles of $u\,dx$ are concentrated at $\{1/x\}$, then, at each place P in K' dividing $\{\psi(x)\}$, $\mathrm{ord}_P((x-a)u) = \mathrm{ord}_P(x-a) + \mathrm{ord}_P u = \mathrm{ord}_P(d(x-a))+1+\mathrm{ord}_P u = \mathrm{ord}_P(dx)+1+\mathrm{ord}_P u = 1+\mathrm{ord}_P(u\,dx) \geq 1$, where a is the value of x at P. Thus $(x-a)u$ is zero at P. Since $(x - a)|\psi(x)$ in $K_0[x]$, it follows that $\psi(x)u$ is zero at P. In particular, $\psi(x)u$ is finite at P. Since this is true for all places P in K' dividing $\{\psi(x)\}$, $\psi(x)u$ is finite at $\{\psi(x)\}$. If $\{\psi(x)\}$ were not [1], then, since $\psi(x)^j u^j$ is zero at P^j for all $j > 0$, there would be an integer j such that $\psi(x)^j u^j$ was zero at $\{\psi(x)\}$. The trace of u can be written in the form $u_1+u_2+\cdots+u_n$ where the u_i are conjugates of u in a function field $L \supset K$ (see §1.25). Each $\psi(x)u_i$ is finite at $\{\psi(x)\}$ and each $\psi(x)^j u_i^j$ would be zero at $\{\psi(x)\}$ if $\{\psi(x)\} \neq [1]$. In this case, $\phi(x)^{nj} = \big(\psi(x)(u_1 + u_2 + \cdots + u_n)\big)^{nj}$ would be a sum of terms, each of which was a product of an element of L finite at $\{\psi(x)\}$ and an element zero at $\{\psi(x)\}$, so $\phi(x)^{nj}$ would be zero at $\{\psi(x)\}$, contrary to the assumption that $\phi(x)$ and $\psi(x)$ are relatively prime. Therefore, $\{\psi(x)\} = [1]$, which implies $\deg \psi = 0$, and $\operatorname{tr} u = \phi/\psi \in K_0[x]$, as was to be shown.

LEMMA 2. *Given any finite set of distinct places* P_1, P_2, $\ldots$, P_μ *in a function field* K, *there is an element* z *of a numerical extension* K' *of* K *such that* $\{1/z\}$ *is a product of distinct places in* K' *including all the given places* P_1, P_2, $\ldots$, P_μ.

PROOF: Let x be a parameter of K. For ν sufficiently large, $\dim(\{1/x\}^{-\nu}) = 1 - \deg(\{1/x\}^{-\nu}) - g = n\nu - g + 1$, where n is the degree of $\{x\}$ in K and g is the genus of K. There are finitely many rational numbers c such that $\{x-c\}$ ramifies (i.e., is divisible by the square of a place in a numerical extension of K) and finitely many such that one of the given places P_i divides $\{x - c\}$. Therefore, there exist constants $c_1, c_2, \ldots, c_\nu$, $c_{\nu+1}$ of K such that the $\{x - c_i\}$ are unramified and relatively prime to $P_1 P_2 \cdots P_\mu$. Let K' be a numerical extension of K in which the $\{x - c_i\}$ are all products of places, and let $y = (x - c_1)(x - c_2) \cdots (x - c_\nu)$. Then $[y] = A\{1/x\}^{-\nu}$, where A is a product of $n\nu$ distinct places in K', say $A = Q_1 Q_2 \cdots Q_{n\nu}$, all distinct from the given places P_i. Finally, let R be a place in K' dividing $\{x - c_{\nu+1}\}$; then R is a place in K' distinct from all the P's and all the Q's.

Let V be the $(n\nu - g + 1)$-dimensional vector space (over the field of constants K_1 of K') of elements of K' divisible by $\{1/x\}^{-\nu}$. The elements of V are finite at Q_i for all i, so evaluation at these $n\nu$ places defines a linear function $V \to K_1^{n\nu}$. An element of the kernel of this linear function is divisible by $A/\{1/x\}^\nu = [y]$. (If $y_1 \in V$, $y_1 \neq 0$, then $\{1/x\}^\nu[y_1]$ is an integral divisor, and if y_1 is zero at Q_i then Q_i divides $\{1/x\}^\nu[y_1]$.) Therefore, if $y_1 \neq 0$ is in the kernel, $[y_1/y]$ is integral, which implies $[y_1/y] = [1]$, that is, y_1 is a constant times y. In other words, the kernel of $V \to K_1^{n\nu}$ is a 1-dimensional subspace generated by y. The image is therefore a subspace of dimension $n\nu - g$. Elementary linear algebra implies that the Q's can be rearranged so that evaluation at the first $n\nu - g$ of the Q's defines an onto linear function $V \to K_1^{n\nu - g}$ whose kernel is the 1-dimensional subspace generated by y. Let this be done, and let $B = Q_1 Q_2 \cdots Q_{n\nu-g}$, $C = A/B = Q_{n\nu-g+1} Q_{n\nu-g+2} \cdots Q_{n\nu}$.

For each $i = 1, 2, \ldots, \mu$, $P_i C/R$ is equivalent to an integral divisor by Abel's theorem (its degree is $1+g-1 = g$), say $[z_i] = R D_i / P_i C$, where $z_i \in K'$ and D_i is an integral divisor in K' of degree g. The order of z_i at P_i is -1, since otherwise P_i would divide D_i (R is distinct from P_i) and C would be equivalent to an integral divisor $R D_i / P_i$ divisible by R and therefore distinct from C, call it E_i; this is impossible, because $C \sim E_i$ would imply $A = BC \sim B E_i$, $[1] \sim A/\{1/x\}^\nu \sim B E_i/\{1/x\}^\nu$, that is, V would contain an element zero at all places in B but nonzero at some place in C, contrary to the choice of B.

The poles of each z_i are concentrated at $P_1 P_2 \cdots P_\mu C$, the order of z_i at P_i is -1, and the order of z_i at P_j is ≥ 0 for $j \neq i$. Therefore, since C is a product of distinct places, $z = z_1 + z_2 + \cdots + z_\mu$ has the desired properties.

LEMMA 3. *Let x be a parameter of a function field K with the property that $\{x\}$ is a product of distinct places, say $\{x\} = P_1 P_2 \cdots P_n$. Given an element u of K, let*

$$u = a_{-N}^{(i)} x^{-N} + \cdots + a_{-1}^{(i)} x^{-1} + \eta_i \qquad (i = 1, 2, \ldots, n)$$

where N is a positive integer independent of i, where the $a_j^{(i)}$ are constants of K, and where η_i has order at least 0 at P_i. Then

$$\mathrm{tr}\, u = \Big(\sum_{i=1}^{n} a_{-N}^{(i)}\Big) x^{-N} + \cdots + \Big(\sum_{i=1}^{n} a_{-1}^{(i)}\Big) x^{-1} + \eta$$

where tr u *is the trace of* u *relative to the extension* $K \supset K_0(x)$
(K_0 *is the field of constants of* K) *and where* η *has order at
least 0 at the place* $\{x\}$ *in* $K_0(x)$.

PROOF: (Dedekind-Weber [**D-W**]) By Theorem 2, there ex-
ists, for each i, an element y_i of K which is integral over $\mathbf{Q}[x]$
and divisible by the restriction to x of $\{x\}^N P_i^{-N}$ but not divisible by the restriction to x of P_i. Then $\mathrm{ord}_P(y_i) \geq N$ for
$P = P_j$ with $j \neq i$, while $\mathrm{ord}_P(y_i) = 0$ for $P = P_i$. A set of
elements $y_1, y_2, \ldots, y_n \in K$ obtained in this way is a basis of
K over $K_0(x)$, as can be shown as follows.

By the theorem of §3.21,

$$n = \deg_K\{x\} = [K\!:\!K_0(x)] \deg_{K_0(x)} x = [K\!:\!K_0(x)].$$

To prove that $y_1, y_2, \ldots, y_n$ give a basis of K over $K_0(x)$ it will
suffice, therefore, to prove that they are linearly independent
over $K_0(x)$, or, what is the same, to prove that if $\psi_1(x), \psi_2(x),$
$\ldots, \psi_n(x) \in K_0[x]$ satisfy $\sum \psi_i(x)y_i = 0$ then $\psi_i(x) = 0$ for all
i. If $\sum \psi_i(x)y_i$ is zero at $\{x\}^s$ for some $s > 0$ then $\psi_i(x)$ must
be zero at $\{x\}^s$ for all i, since otherwise it would be possible to
reorder the y's and ψ's and choose a nonnegative integer $t < s$
such that each $\psi(x)/x^t$ was in $K_0[x]$ and the constant term
of $\psi_1(x)/x^t$ was nonzero; this would imply a contradiction,
because $\sum \psi_i(x)x^{-t}y_i$ would be zero at $\{x\}$ ($s > t$), while
at P_1 the first term of the sum would be nonzero (neither
$\psi_1(x)x^{-t}$ nor y_1 is zero at P_1) and the other $n-1$ terms would
all be zero at P_1 ($\psi_i(x)x^{-t}$ is finite at P_1 and y_i is zero there
for $i > 1$). In particular, if $\sum \psi_i(x)y_i = 0$ then $\psi_i(x)$ is zero
at $\{x\}^s$ for all i and all s, which shows $\psi_i(x) = 0$ for all i.

Therefore, the equations $uy_i = \sum v_{ij}y_j$ define elements
v_{ij} of $K_0(x)$, and tr $u = \sum_{i=1}^n v_{ii}$. Since $x^N u$ and y_i are
finite at all the P_i, $\sum x^N v_{ij}y_j$ is finite at all the P_i. If
$q_i(x) \in K_0[x]$ is a common denominator of $v_{i1}, v_{i2}, \ldots, v_{in}$, say
$v_{ij} = \psi_{ij}(x)/q_i(x)$ where $\psi_{ij}(x) \in K_0[x]$, then $q_i(x)x^N uy_i =$
$\sum_j \psi_{ij}(x)x^N y_j$ is divisible by $\{x\}$ at least as many times as

$q_i(x)$ is, which implies that $\psi_{ij}(x)x^N$ is divisible by $\{x\}$ at least as many times as $q_i(x)$ is (as was just shown), which implies $v_{ij}x^N = \psi_{ij}(x)x^N/q_i(x)$ is finite at $\{x\}$ and therefore that $x^N uy_i$ is finite at $\{x\}$. On the other hand, $x^{-N}y_j$ is finite at P_k whenever $k \neq j$, so $v_{ij}y_j = v_{ij}x^N x^{-N}y_j$ is finite at P_k whenever $k \neq j$. When $k \neq i$, $uy_i = ux^N x^{-N}y_i$ is finite at P_k, so every term in the equation $uy_i = \sum v_{ij}y_j$ is finite at P_k with the possible exception of the term $v_{ik}y_k$. Therefore, this term is finite at P_k, that is, $v_{ik}y_k$ is finite at P_k when $i \neq k$. Since y_k has order 0 at P_k, this shows that v_{ik} is finite at P_k when $i \neq k$. Since $v_{ik} \in K_0(x)$, v_{ik} is finite at $\{x\}$ when $i \neq k$. Thus,

$$(v_{ii} - u)y_i = -\sum_{j \neq i} v_{ij}y_j$$

is finite at P_i because all terms on the right are finite at P_i (v_{ij} is finite at $\{x\}$ and y_j is finite—in fact zero—at P_i). Since y_i is of order 0 at P_i, it follows that $v_{ii} - u$ is finite at P_i. Therefore,

$$\zeta_i = (v_{ii} - u) + (u - a^{(i)}_{-N}x^{-N} - \cdots - a^{(i)}_{-1}x^{-1}) = (v_{ii} - u) + \eta_i$$

is finite at P_i. Thus $v_{ii} = a^{(i)}_{-N}x^{-N} + \cdots + a^{(i)}_{-1}x^{-1} + \zeta_i$ where ζ_i is in $K_0(x)$ (all other terms of the equation are in $K_0(x)$). Therefore, ζ_i is in fact finite at $\{x\}$. In conclusion, then, $\operatorname{tr} u = \sum v_{ii} = (\sum a^{(i)}_{-N})x^{-N} + \cdots + (\sum a^{(i)}_{-1})x^{-1} + \zeta$, where $\zeta = \sum \zeta_i$ is finite at $\{x\}$, as was to be shown.

PROOF OF THE THEOREM: By Lemma 2, there is an x in a numerical extension K'' of K' such that $\{1/x\}$ is a product of distinct places in K'' including all places in K'' which divide B. Let $u \in K''$ be such that $u\,dx$ represents the given differential. For each place P_i dividing $\{1/x\}$, x^{-1} is a local parameter at P_i and the residue of $u\,dx = u(-x^2)d(x^{-1})$ at P_i is the coefficient of $x = (x^{-1})^{-1}$ in the expansion of $-x^2u$ in powers of x^{-1} at P_i. When constants $a^{(i)}_j$ in K'' are defined by $u = a^{(i)}_N x^N +$

$a_{N-1}^{(i)} x^{N-1} + \cdots + a_0^{(i)} + a_{-1}^{(i)} x^{-1} + \eta_i$, where η_i has order greater than 1 at P_i, the residue of $u\,dx$ at P_i is $-a_{-1}^{(i)}$. By Lemma 3, $\mathrm{tr}(-x^2 u) = (-\sum a_N^{(i)}) x^{N+2} + \cdots + (-\sum a_{-1}^{(i)}) x + \eta$, where η has order at least 0 at $\{x^{-1}\}$, is the expansion of $\mathrm{tr}(-x^2 u)$ in powers of the local parameter x^{-1} at $\{x^{-1}\}$. By Lemma 1, $\mathrm{tr}\,u$ is a polynomial in x. Therefore, $\mathrm{tr}(-x^2 u)$ is a polynomial in x with no terms of degree 0 or 1. By the uniqueness of expansions in the form $b_{N+2} x^{N+2} + \cdots + b_1 x + \eta$, it follows that $\eta = 0$, $\sum_i a_{-1}^{(i)} = 0$, and $-\sum_i a_j^{(i)}$ is the coefficient of x^j in $\mathrm{tr}\,u$ for $j \geq 0$. Since $a_{-1}^{(i)}$ is the residue of $u\,dx$ at P_i in K' as well as in K'', $\sum a_{-1}^{(i)} = 0$ is the equation that was to be proved.

§A.5 Holomorphic Differentials.

A differential in a function field K is *holomorphic* if it has no poles, that is, if its order at any place in any numerical extension of K is nonnegative.

The holomorphic differentials in K obviously form a vector space over the field of constants of K.

THEOREM. *The dimension of the holomorphic differentials in a function field K as a vector space over the field of constants of K is equal to the genus of K.*

The proof, which will be given in §A.10, will include a construction, following [D-W], of the holomorphic differentials in K.

§A.6 Integral Bases.

PROPOSITION. *Let x be a parameter of a function field K. There is a basis $y_1, y_2, \ldots, y_n$ of K over $K_0(x)$, where K_0 is the field of constants of K, such that an element z of K is integral over $\mathbf{Q}[x]$ (or, what is the same, integral over $K_0[x]$) if and only if its unique representation in the form $z = v_1 y_1 + v_2 y_2 + \cdots + v_n y_n$ with $v_i \in K_0(x)$ has $v_i \in K_0[x]$ for all i.*

DEFINITION. Such a basis of K over $K_0(x)$ is an *integral basis* with respect to x.

PROOF: The proof is a simple adaptation of the proof of the analogous fact in §2.6. Because $[K:\mathbf{Q}(x)] < \infty$, $K = \mathbf{Q}(x, y)$ for some $y \in K$ integral over $\mathbf{Q}[x]$. By the second corollary of §1.28, there is a $\Delta \in \mathbf{Q}[x]$ such that every element of K integral over $\mathbf{Q}[x]$ is of the form $(a_0 + a_1 y + a_2 y^2 + \cdots + a_{m-1} y^{m-1})/\Delta$, where the a_i are in $\mathbf{Q}[x]$ and $m = [K:\mathbf{Q}(x)]$. Each a_i can be written in the form $a_i = q_i \Delta + r_i$, where $\deg r_i < \deg \Delta$. Thus, every element of K integral over $\mathbf{Q}[x]$ is of the form $q_0 + q_1 y + q_2 y^2 + \cdots + q_{m-1} y^{m-1} + \delta$, where δ is of the form $(\sum_{i=0}^{m-1} s_i y^i)/\Delta$, with $s_i \in \mathbf{Q}[x]$, $\deg s_i < \deg \Delta$, and where δ is integral over $\mathbf{Q}[x]$. As was noted in §3.24, integrality of $\sum s_i y^i/\Delta$ imposes homogeneous linear conditions on the $m \cdot \deg \Delta$ coefficients of the s_i, so there is a finite set δ_1, δ_2, ..., δ_k of such δ's with the property that every such δ is a linear combination of these with coefficients in $\mathbf{Q}$. Therefore 1, y, y^2, ..., y^{m-1}, δ_1, δ_2, ..., δ_k span the set of all elements of K integral over $\mathbf{Q}[x]$ as a module over $\mathbf{Q}[x]$.

Since, *a fortiori*, it spans this set over $K_0[x]$, the algorithm of §2.6 can be used to find a basis of K over $K_0(x)$ whose elements span the same $K_0[x]$–module of all elements of K integral over $\mathbf{Q}[x]$. Specifically, at step one the spanning set is rearranged to make the discriminant (note that $K_0[x]$ is a natural ring) of the first $n = [K:K_0(x)]$ elements be a nonzero element of $K_0[x]$ of minimum degree. At step two, all following elements are reduced by subtracting multiples of the first n to make the coefficients of their representations in this basis *proper* rational functions (with a numerator of smaller degree than the denominator) or zero. Repetition of these steps *reduces* the degree of the discriminant of the first n elements of the spanning set, unless all elements that follow the first n are zero.

§A.7 Normal Bases.

Let x be a parameter of a function field K. The *order at* $x = \infty$ of a nonzero element z of K integral over $\mathbf{Q}[x]$ is the least integer e for which z/x^e is finite at $\{1/x\}$.

PROPOSITION. *Given a parameter x of a function field K, there is an integral basis y_1, y_2, $\ldots$, y_n of K with respect to x such that the order at $x = \infty$ of any nonzero element z of K integral over $\mathbf{Q}[x]$ is equal to the maximum order of the terms in its expansion $z = \sum c_{ij} x^i y_j$, where the c_{ij} are nonzero elements of K_0.*

DEFINITION: Such an integral basis is a *normal basis* of K with respect to x.

PROOF: Let y_1, y_2, $\ldots$, y_n be an integral basis of K with respect to x. Let e_i be the order of y_i at $x = \infty$. Since $y_i/x_i^{e_i}$ is finite at $\{1/x\}$, it represents an element of the ring of values of K at $\{1/x\}$. Let V denote the ring of values of K at $\{1/x\}$. If the n elements of V represented by y_i/x^{e_i} for $i = 1, 2, \ldots, n$ are not linearly independent over K_0 (V is a ring containing K_0 in a natural way, and is therefore a vector space over K_0) then there exist constants $b_1, b_2, \ldots, b_n \in K_0$ not all zero such that $\sum_{i=1}^{n} b_i y_i x^{-e_i}$ is zero at $\{1/x\}$. Let j be an integer, $1 \leq j \leq n$, such that $b_j \neq 0$ and e_j is as large as possible. Since $x \sum_{i=1}^{n} b_i y_i x^{-e_i}$ is finite at $\{1/x\}$, the order of $y_j' = \sum_{i=1}^{n} b_i y_i x^{e_j - e_i}$ is at most $e_j - 1$. Thus, replacement of y_j by y_j' gives a new integral basis of K with respect to x ($e_j - e_i \geq 0$ and $b_j \neq 0$) in which $\sum e_i$ is reduced by at least 1. Therefore, a finite number of repetitions of these steps gives an integral basis y_1, y_2, $\ldots$, y_n of K with respect to x for which the elements of V represented by y_i/x^{e_i} are linearly independent over K_0. It will be shown that such an integral basis is a normal basis.

Let $z = \sum c_{ij} x^i y_j$ where the sum contains at least one term, where the i's are ≥ 0, and where the $c_{ij} \in K_0$ are nonzero. Let

e be the maximum over all terms of this sum of $i + e_j$. It is to be shown that e is the order of z. Since $x^i y_j / x^e$ is zero at $\{1/x\}$ when $i + e_j < e$, the element of V represented by z/x^e is the same as the element of V represented by $\sum c_{ij} x^i y_j / x^e = \sum c_{ij} y_j / x^{e_j}$, where the sum is over all terms (there is at most one term for each $j = 1, 2, \ldots, n$) in which $i + e_j = e$. Thus, since this is a linear combination of linearly independent elements in which the coefficients are not all zero, the element of V represented by z/x^e is not zero. Therefore, e is the order of z at $x = \infty$, as was to be shown.

COROLLARY. *The genus of K is $1 + \sum(e_i - 1)$.*

DEDUCTION: For ν sufficiently large, the dimension of the subspace of K over K_0 containing elements divisible by $\{1/x\}^{-\nu}$ is $n\nu - g + 1$, where $n = [K : K_0(x)]$ and g is the genus of K (§3.26). But an element z of K is divisible by $\{1/x\}^{-\nu}$ if and only if it is integral over $\mathbf{Q}[x]$ and of order at most ν at $x = \infty$. Thus, z is in this subspace if and only if its representation relative to the normal basis $y_1, y_2, \ldots, y_n$ is of the form $z = \sum \phi_j(x) y_j$, where $\phi_j(x) \in K_0[x]$ is of degree $\nu - e_j$ at most. Therefore, $n\nu - g + 1 = \sum_{i=1}^{n}(\nu - e_i + 1)$, which gives the desired formula for g.

§A.8. The Dual of a Normal Basis.

Let x be a parameter of a function field K. The trace relative to the field extension $K \supset K_0(x)$ (where K_0 is the field of constants of K) is a linear form $K \to K_0(x)$. Therefore, the trace of the product is a symmetric bilinear form $K \times K \to K_0(x)$. This bilinear form is nondegenerate, because if $f \in K$ were nonzero and satisfied $\mathrm{tr}_{K/K_0(x)}\, fg = 0$ for all $g \in K$, then $\mathrm{tr}_{K/K_0(x)}\, 1$ would be 0 (set $g = f^{-1}$), contrary to the fact that $\mathrm{tr}_{K/K_0(x)}\, 1 = [K : K_0(x)]$. Given any basis $y_1, y_2, \ldots, y_n$ of K over $K_0(x)$, one can therefore define a *dual basis* $z_1, z_2, \ldots,$

z_n of K over $K_0(x)$ by the equations

$$\mathrm{tr}_{K/K_0(x)}\, y_i z_j = \begin{cases} 1 & \text{if } i = j \\ 0 & \text{otherwise.} \end{cases}$$

PROPOSITION. *Let $y_1, y_2, \ldots, y_n$ be a normal basis of K with respect to x, and let e_i be the order of y_i at $x = \infty$ for $i = 1, 2, \ldots, n$. Then $v_i = y_i/x^{e_i}$ for $i = 1, 2, \ldots, n$ defines a normal basis of K with respect to $1/x$, and the order of v_i at $(1/x) = \infty$ is e_i.*

PROOF: The poles of v_i are concentrated at $\{x\}\{1/x\}$ (the poles of y_i are concentrated at $\{1/x\}$ and the poles of $(1/x)^{e_i}$ are concentrated at $\{x\}$). By the definition of e_i, v_i has no poles at $\{1/x\}$. Therefore, the poles of v_i are concentrated at $\{x\}$, that is, v_i is integral over $\mathbf{Q}[1/x]$.

Let $w \in K$ be nonzero and integral over $\mathbf{Q}[1/x]$ and let e' be its order at $(1/x) = \infty$. Then (as was just shown in the case of y_i) $w/(1/x)^{e'} = x^{e'}w$ is integral over $\mathbf{Q}[x]$. Thus $x^{e'}w = \sum c_{ij} x^i y_j = \sum c_{ij}(1/x)^{-i-e_j} v_j$. Let e be the order of $x^{e'}w$ at $x = \infty$. Then $e \leq e'$, so the exponents of $1/x$ in the expansion $w = \sum c_{ij}(1/x)^{e'-i-e_j} v_j$ are all nonnegative. Therefore, $v_1, v_2, \ldots, v_n$ is an integral basis of K with respect to $1/x$.

Since $v_j/(1/x)^{e_j} = y_j$ is finite at $\{x\}$, the order of v_j at $(1/x) = \infty$ is at most e_j. Since $x^{-1}y_j$ is not finite at $\{x\}$ (if it were, $x^{-1}y_j$ would be integral over $\mathbf{Q}[x]$, contrary to the assumption that $y_1, y_2, \ldots, y_n$ is an integral basis of K with respect to x), e_j is the order of v_j at $(1/x) = \infty$. Therefore, the order of $c_{ij}(1/x)^{e'-i-e_j} v_j$ at $(1/x) = \infty$ is $e' - i$. If i were positive in all terms of $x^{e'}w = \sum c_{ij} x^i y_j$, then (the y_i are an integral basis) $x^{e'-1}w$ would be finite at $\{x\}$, contrary to the definition of e'. Therefore $i = 0$ in at least one term, so $w = \sum c_{ij}(1/x)^{e'-i-e_j} v_j$ contains at least one term of order e' at $(1/x) = \infty$, which shows that $v_1, v_2, \ldots, v_n$ is a normal basis of K with respect to $1/x$.

§A.9.

PROPOSITION 1. *Let y_1, y_2, $\ldots$, y_n be an integral basis of a function field K with respect to a parameter x of K, and let z_1, z_2, $\ldots$, z_n be its dual basis. Let D_x be the different of y_1, y_2, $\ldots$, y_n with respect to K as an extension of the natural ring $K_0[x]$. Then $D_x[z_i]_x$ is integral for all $i = 1, 2, \ldots, n$.*

(Here $[z_i]_x$ is the restriction to x of the global divisor $[z_i]$.)

PROOF: In the notation of §1.28, $D_x = [F'(\bar{\alpha})]$, where $F'(\bar{\alpha})$ is as defined in §1.26 with $r = K_0[x]$ and $\alpha_i = y_i$ ($i = 1, 2, \ldots, n$). Since z_i has the property that $\mathrm{tr}_{K/K_0(x)}\, y z_i \in K_0[x]$ whenever y is integral over $K_0[x]$, $\mathrm{tr}_{K/K_0(x)}\big(z_i q(X)\big) = \phi(X)$ has coefficients in $K_0[x]$, where the notation of §1.28 is modified by setting X in place of x so that $q(X) = F(X)/(X - \bar{\alpha})$. As in §1.28, $\phi(\bar{\alpha}) = z_i F'(\bar{\alpha})$. Thus, $[z_i F'(\bar{\alpha})]_x$ is integral over $K_0[x]$, as was to be shown.

PROPOSITION 2. (Cf. §2.8.) *Let D_x be as above, let P be a place in K at which x is finite, and let a be the value of x at P. Then the numerator of $[x - a]_x/D_x$ is divisible by P_x. In fact, if e is the multiplicity with which P_x divides $[x-a]_x$, then $e - 1$ is the multiplicity with which P_x divides D_x.*

PROOF: Assume, without loss of generality, that $a = 0$ and that $\{x\}$ is a product of places, say $\{x\} = P_1^{e_1} P_2^{e_2} \cdots P_k^{e_k}$. Let V be the ring of values of K at $[x]_x$, and let ι be the natural map of elements of K finite at $[x]_x$ to V. Then K_0 is included in V in a natural way and $\iota(y_1), \iota(y_2), \ldots, \iota(y_n)$ are linearly independent over K_0 because $a_1 \iota(y_1) + a_2 \iota(y_2) + \cdots + a_n \iota(y_n) = 0$ for $a_i \in K_0$ implies $(a_1 y_1 + a_2 y_2 + \cdots + a_n y_n)/x$ is integral over $\mathbf{Q}[x]$, which in turn implies, because the y's are an integral basis with respect to x, that the a_i are all 0. Therefore, as in Lemma 1 of §2.7, the $\iota(y_i)$ are a basis of V over K_0, K is p-regular for $p = x$, and, by Proposition 1 of §1.32, the image in $V[X, u_1, u_2, \ldots, u_n]$ of $F(X) = N_{K/K_0(x)}(X - u_1 y_1 - u_2 y_2 - \cdots - u_n y_n)$ is $\prod_{i=1}^{k} N_i(X - u_1 y_1 - \cdots - u_n y_n)^{e_i}$ where N_i is the norm with respect to the field extension $V_{(P_i)_x} \supset K_0$ of $X - u_1 y_1 - \cdots - u_n y_n$ when the y_i are regarded as representing elements of $V_{(P_i)_x}$. (The natural ring $K_0[x] \bmod x$ is the field K_0, so $k_p = K_0$ in this case.) Because P_i is a place, every element of K finite at $\{x\}$ has a value in K_0 at P_i, so $V_{(P_i)_x} = K_0$. Thus, the image of $F(X)$, which has coefficients in $K_0(x)$ so that its image in $V[X, u_1, u_2, \ldots, u_n]$ has coefficients in K_0, is simply $\prod_{i=1}^{k}(X - \theta_1^{(i)} u_1 - \theta_2^{(i)} u_2 - \cdots - \theta_k^{(i)} u_k)^{e_i}$, where $\theta_j^{(i)} \in K_0$ is the value of y_j at P_i.

It follows that $F'(X) \equiv \sum_{i=1}^{k} e_i(X - \theta_1^{(i)} u_1 - \cdots - \theta_n^{(i)} u_n)^{-1} F(X) \bmod x$. D_x is the divisor in K as an extension of $K_0[x]$ represented by the polynomial obtained by setting $y_1 u_1 + y_2 u_2 + \cdots + y_n u_n$ in place of X in $F'(X)$. Since $P_1^{e_1}$ divides $[x]_x$ and since all terms of $F'(y_1 u_1 + \cdots + y_n u_n) \bmod x$ but the first are obviously divisible by $P_1^{e_1}$, the statement to be proved is that the divisor represented by

(1)
$$e_1\big((y_1 - \theta_1^{(1)})u_1 + (y_2 - \theta_2^{(1)})u_2 + \cdots + (y_n - \theta_n^{(1)})u_n\big)^{e_1-1}.$$
$$\prod_{i>1}\big((y_1 - \theta_1^{(i)})u_1 + \cdots + (y_n - \theta_n^{(i)})u_n\big)^{e_i}$$

is not divisible by $(P_1)_x^{e_1}$. (It is obviously divisible by $(P_1)_x^{e_1-1}$.)

Since P_i is a place, $(P_i)_x = [x, z]_x$ for some $z \in K$ integral over $K_0[x]$. Since $P_i \neq P_1$, z is not zero at P_1. Since the y's are an integral basis, $z = \sum \phi_j(x) y_j$ and the value at P_i of z is $\sum \phi_j(0)\theta_j^{(i)} = 0$. If $(P_1)_x$ divided $[\sum_j (y_j - \theta_j^{(i)})u_i]_x$, the value of z at P_1 would be $\sum \phi_j(0)\theta_j^{(i)} = 0$. Thus, $(P_1)_x$ does not divide the factors of (1) other than $[\sum (y_j - \theta_j^{(1)})u_j]_x^{e_1-1}$. If $e_1 = 1$, the proposition follows. If $e_1 > 1$, it remains to show that $(P_1)_x^2$ does not divide $[\sum (y_j - \theta_j^{(1)})u_j]_x$.

If $(P_1)_x^2$ did divide this divisor, then each y_j would be of the form $\theta_j^{(1)} + v_j$, where $\operatorname{ord}_{P_1}(v_j) \geq 2$. Since $\operatorname{ord}_{P_1}(x) = e_1 \geq 2$

by assumption, any element of K integral over $K_0[x]$ would be of the form $z = \sum \phi_j(x)y_j = \beta + w$, where $\beta \in K_0$ and $\mathrm{ord}_{P_1}(w) \geq 2$. But then $(P_1)_x^2$ would divide $[x, z]_x$ whenever z was integral over $K_0[x]$ and zero at P_1, contrary to the fact that $(P_1)_x$ can be expressed in the form $[x, z]_x$.

§A.10. Construction of Holomorphic Differentials.

THEOREM. *Let* y_1, y_2, ..., y_n *be a normal basis of* K *with respect to* x *and let* z_1, z_2, ..., z_n *be its dual basis. A differential* ω *in* K *is holomorphic if and only if the element* $u \in K$ *for which* $\omega = u\,dx$ *is of the form* $u = \psi_1 z_1 + \psi_2 z_2 + \cdots + \psi_n z_n$, *where the* ψ_i *are in* $K_0[x]$ *and their degrees satisfy* $\deg \psi_i \leq e_i - 2$, *where* e_i *is the order of* y_i *at* $x = \infty$.

PROOF: If ϕ_1, ϕ_2, ..., $\phi_n \in K_0[x]$, then

$$\sum_{i=1}^{n} \phi_i\psi_i = \sum_{i=1}^{n}\sum_{j=1}^{n} \phi_i\psi_j \, \mathrm{tr}(y_i z_j) = \mathrm{tr}(fu),$$

where $f = \sum_{i=1}^{n} \phi_i y_i$. If $u\,dx$ is holomorphic, then, since f is integral over $\mathbf{Q}[x]$, $fu\,dx$ has poles concentrated at $\{1/x\}$, and $\sum \phi_i\psi_i \in K_0[x]$ by Lemma 1 of §A.4. Therefore, since the $\phi_i \in K_0[x]$ are arbitrary, the ψ_i must all be in $K_0[x]$.

Let $v_i = y_i/x^{e_i}$ and $w_i = x^{e_i}z_i$ for $i = 1, 2, \ldots, n$. Then v_i is a normal basis of K with respect to $1/x$ and w_i is its dual basis. If $\omega = u\,dx = -x^2 u\,d(1/x)$ is holomorphic, then $fx^2 u\,d(1/x)$ has poles only at $\{x\}$ whenever $f = \sum \theta_i v_i$ with $\theta_i \in K_0[1/x]$. Therefore, $\mathrm{tr}(fx^2 u) = \sum \theta_i x^2 \psi_i x^{-e_i}$ is in $K_0[1/x]$, which is to say that, for each i, $\psi_i(x)/x^{e_i-2}$ is a polynomial in x^{-1}, or, what is the same, that the degree of ψ_i is at most $e_i - 2$.

Thus, a holomorphic differential is necessarily of the form $(\sum \psi_i z_i)dx$, where $\psi_i \in K_0[x]$ and $\deg \psi_i \leq e_i - 2$. It remains to show that all these differentials are indeed holomorphic, that is, that $x^i z_j dx$ is holomorphic whenever $0 \leq i \leq e_j - 2$.

At any place P in K or a numerical extension of K at which x is finite, say x is a at P, $\mathrm{ord}_P(z_j dx) = \mathrm{ord}_P(z_j) + \mathrm{ord}_P(dx) =$

$\text{ord}_P(z_j) + \text{ord}_P(x-a) - 1$. Let $e = \text{ord}_P(x-a)$. Now $[x - a]_x[z_j]_x = (D_x[z_j]_x)([x-a]_x/D_x)$. The first factor is integral by Proposition 1 of §A.9 and the second factor has numerator divisible by P_x by Proposition 2. Therefore $\text{ord}_P((x-a)z_j) \geq 1$ and $\text{ord}_P(z_j dx) \geq 0$. Thus, for any $i \geq 0$ and for any $j = 1$, $2, \ldots, n$, $x^i z_j dx$ has poles concentrated at $\{1/x\}$.

When x is replaced by $1/x$ and y_j by $v_j = y_j/x^{e_j}$, which replaces z_j by $w_j = x^{e_j} z_j$ and $x^i z_j dx$ by $(1/x)^i w_j d(1/x) = -z_j x^{e_j - 2 - i} dx$, it follows that the poles of $-z_j x^{e_j - 2 - i} dx$ are concentrated at $\{x\}$ for $e_j - 2 - i \geq 0$ and $j = 1, 2, \ldots, n$. Thus, if $0 \leq i \leq e_j - 2$, the poles of $x^i z_j dx$ are concentrated both at $\{1/x\}$ and at $\{x\}$, so $x^i z_j dx$ is holomorphic, as was to be shown.

COROLLARY. (Theorem of §A.5) *The genus of K is the dimension of the vector space of holomorphic differentials in K.*

DEDUCTION: If $e_j > 0$, the vector space of polynomials of degree i, $0 \leq i \leq e_j - 2$, has dimension $e_j - 1$. If $e_j = 0$, it has dimension $0 = (e_j - 1) + 1$. A normal basis contains exactly one constant element, so exactly one e_j is 0, and the dimension of the space of holomorphic differentials is $\sum(e_j - 1) + 1 = g$ (§A.7).

§A.11. Examples.

If K is the function field of a hyperelliptic curve $y^2 = f(x)$, where $f \in \mathbf{Q}[x]$ is of degree $2d$ with distinct roots, then 1, y is a normal basis. Since $e_1 = 0$ and $e_2 = d$, the genus is $((-1) + (d-1)) + 1 = d - 1$. Since the 2×2 symmetric matrix which represents the trace of the product relative to this basis is the diagonal matrix whose entries on the diagonal are 2 and $2f(x)$, the dual basis is $\frac{1}{2}$, $\frac{y}{2f(x)} = \frac{1}{2y}$, so the holomorphic differentials are those of the form $g(x)\, dx/y$, where $g(x)$ is a polynomial of degree $\leq d - 2$.

If K is the function field of the Klein curve $y^3 + x^3 y + x = 0$, then, as is shown in §3.28, $g = 3$. The holomorphic differentials in K can be found as follows.

At any place $P = \{x-a, y-b\}$ in K or a numerical extension of K at which x, and therefore y, is finite, either $x - a$ or $y - b$ is a local parameter. (Differentiation of $y^3 + x^3y + x = 0$ gives $(3y^2 + x^3)dy + (3x^2y + 1)dx = 0$, so $x - a$ is a local parameter at P if and only if $F_y = 3y^2 + x^3$ is nonzero at P, and $y - b$ is a local parameter at P if and only if $F_x = 3x^2y+1$ is nonzero at P. If F_x and F_y were both zero at P, then multiplication of $F_y = 0$ by $3x^4$ would give $(3x^2y)^2 + 3x^4x^3 = 0$, and therefore, since $3x^2y = -1$ at P, $1 + 3x^7 = 0$ at P. But multiplication of $y^3 + x^3y + x = 0$ by $27x^6$ would give $(3x^2y)^3+9x^4x^3(3x^2y)+27x^6x = 0$, $(3x^2y)^3+9x^7(3x^2y+3) = 0$, and therefore $-1 + 9x^7 \cdot 2 = 0$ at P, which would contradict $1 + 3x^7 = 0$ at P.)

Let $u\,dx$ be a holomorphic differential in K. If $x - a$ is a local parameter at P, then u is finite at P. If $x - a$ is not a local parameter at P, then $y - b$ is a local parameter at P. Since $u\,dx = -uF_ydy/F_x$, it follows that uF_y/F_x is finite at P. Since F_x is finite at P, uF_y is finite at P. Since F_y is finite at any place where x is finite, uF_y *is finite at any place P in K or a numerical extension of K where x is finite.* Therefore, uF_y is integral over $\mathbf{Q}[x]$.

Conversely, if $uF_y = v$ is integral over $\mathbf{Q}[x]$, then, at any place P where x is finite, either $x - a$ is a local parameter, in which case $F_y \neq 0$ and $u = v/F_y$ is finite at P so $u\,dx$ has no pole at P, or $y - b$ is a local parameter at P, in which case $F_x \neq 0$ and $u\,dx = -uF_ydy/F_x$ has no pole at P. Thus, the determination of the holomorphic differentials in K reduces to the determination of those elements v of K integral over $\mathbf{Q}[x]$ for which $(v/F_y)dx$ has no poles at places where $x = \infty$. (The same is true of the determination of the holomorphic differentials in the function field K of any curve $F(x,y) = 0$ monic in y for which F_x and F_y do not vanish simultaneously for finite x.)

The two places where $x = \infty$ are $Q = \{1/y, x/y\}$ and $R = \{y/x, 1/x\}$. At Q, $s = x/y$ is a local parameter, $t = 1/y$ has a double zero, and $s^3 + t^3 s + t = 0$ holds. Then $t = -s^3 - st^3 = -s^3 + s^{10} + \cdots$, $x = s/t = -s^{-2} + \cdots$, $dx = (2s^{-3} + \cdots)ds$, $F_y = 3y^2 + x^3 = 3t^{-2} + (s/t)^3 = t^{-3}(3t + s^3) = (-s^3 + \cdots)^{-3}(-2s^3 + \cdots) = 2s^{-6} + \cdots$, so $v\, dx/F_y = v(s^3 + \cdots)ds$, and $v\, dx/F_y = u\, dx$ has a pole at Q if and only if v has a pole at Q of order greater than 3. At R, $z = 1/x$ is a local parameter, $w = y/x$ has a double zero, and $z^3 + w^3 z + w = 0$. Then $dx = -z^{-2}dz$ and $F_y = 3(w/z)^2 + (1/z)^3 = z^{-3}(1 + 3zw^2) = z^{-3} + \cdots$, so $v\, dx/F_y = v(-z + \cdots)dz$ has a pole at R if and only if v has a pole at R of order greater than 1.

As was seen in §3.28, $v = \xi_0 + \xi_1 y + \xi_2 y^2$ where $\xi_i \in \mathbf{Q}[x]$. It is easy to check that when $v = a + bx + cy$ for constants a, b, and c, the pole of v at R has order ≤ 1 and its pole at Q has order ≤ 3. Since $g = 3$, it follows that the most general holomorphic differential in K is

$$\frac{a + bx + cy}{3y^2 + x^3}\, dx$$

where a, b, and c are constants.

§A.12.   The Riemann-Roch Theorem for Integral Divisors.

As in §A.1, let $P_1, P_2, \ldots, P_k$ be places in a function field K, let u_i be a local parameter at P_i for $i = 1, 2, \ldots, k$, let $z \in K$ have poles concentrated at $P_1 P_2 \cdots P_k$, and let $(\phi_1, \phi_2, \ldots, \phi_k)$ be the principal parts of z relative to the parameters u_i, that is, let $\phi_i(X)$ be the unique polynomial with coefficients in K_0 (the field of constants of K) and with zero constant term for which $z - \phi_i(u_i^{-1})$ is finite at P_i $(i = 1, 2, \ldots, k)$.

If $w_1, w_2, \ldots, w_g$ are a basis of the vector space (over K_0) of holomorphic differentials in K, the statement that the sum of the residues of zw_i is zero gives g linear conditions satisfied

by the principal parts $(\phi_1, \phi_2, \ldots, \phi_k)$ of z. Specifically, let
N be an integer large enough that $N \geq \deg \phi_i$ for all i, let
polynomials $\xi_i^{(j)}$ of degree $N-1$ at most with coefficients in K_0
be defined by the conditions $\omega_j = (\xi_i^{(j)}(u_i) + o(u_i^N)) du_i$, and
let $\langle \phi_i, \xi_i^{(j)} \rangle$ denote the coefficient of X^{-1} in $\phi_i(X^{-1}) \xi_i^{(j)}(X)$.
Then $\langle \phi_i, \xi_i^{(j)} \rangle$ is the residue of $z\omega_i$ at P_i and the principal
parts $(\phi_1, \phi_2, \ldots, \phi_k)$ of z satisfy the g homogeneous linear
conditions $\sum_{i=1}^{k} \langle \phi_i, \xi_i^{(j)} \rangle = 0$ $(j = 1, 2, \ldots, g)$. The main
theorem is that *these necessary conditions are sufficient.* That
is, given a k-tuple $(\phi_1, \phi_2, \ldots, \phi_k)$ of polynomials which satisfy
$\sum_{i=1}^{k} \langle \phi_i, \xi_i^{(j)} \rangle = 0$ $(j = 1, 2, \ldots, g)$ when the polynomials
$\xi_i^{(j)}$ are defined as above, relative to some N greater than the
degrees of all the ϕ_i, there is an element $z \in K$ with poles
concentrated at $P_1 P_2 \cdots P_k$ whose principal parts, relative to
the parameters u_i, are $(\phi_1, \phi_2, \ldots, \phi_k)$.

PROOF: Increasing N does not alter the conditions satisfied
by $(\phi_1, \phi_2, \ldots, \phi_k)$. Therefore, one can assume without
loss of generality that, with $A = P_1 P_2 \cdots P_k$, $\dim_K(A^{-N}) +$
$\deg_K(A^{-N}) + g = 1$ (§3.26). Thus, $\dim_K(A^{-N}) = kN - g + 1$.
Since the principal parts of an element of K are zero if and only
if the element is constant, the principal parts of elements of K
divisible by A^{-N} are a vector space of dimension $kN - g$ over
K_0. The principal parts of such a function are described by a
k-tuple $(\phi_1, \phi_2, \ldots, \phi_k)$ of polynomials of degree N, at most,
with coefficients in K_0 and with constant term zero. Since
the space of all such k-tuples is a vector space of dimension
kN over K_0, the principal parts of elements of K divisible by
A^{-N} are a subspace of codimension g. Elements of the sub-
space satisfy the g conditions $\sum_{i=1}^{k} \langle \phi_i, \xi_i^{(j)} \rangle = 0$ $(j = 1, 2,$
$\ldots, g)$. These conditions are *independent* for N sufficiently
large, because, for any fixed i, if $a_1 \omega_1 + a_2 \omega_2 + \cdots + a_g \omega_g \neq 0$
then $a_1 \xi_i^{(1)} + a_2 \xi_i^{(2)} + \cdots + a_g \xi_i^{(g)} \neq 0$ for N large. These
conditions therefore define a subspace of codimension g which

contains the principal parts of elements of K divisible by A^{-N} and therefore, because it has the same dimension, *coincides* with the principal parts. Therefore, the necessary conditions are sufficient, as was to be shown.

COROLLARY. (Riemann-Roch Theorem for Integral Divisors) *Let A be an integral divisor in a function field K. Then $\dim_K(A^{-1}) + \deg_K(A^{-1}) + g = 1 + \mu$ where μ is the dimension of the space of holomorphic differentials zero at A. More precisely, μ is the dimension of the vector space of differentials ω which have the property that, for any place P in any numerical extension of K, $\text{ord}_P(\omega)$ is at least as large as the number of times P divides A.*

DEDUCTION: Taking a numerical extension of K does not change $\dim_K(A^{-1})$, $\deg_K(A^{-1})$, g, or μ (see §3.23). Therefore, one can assume without loss of generality that A is a product of powers of places, say $A = P_1^{e_1} P_2^{e_2} \cdots P_k^{e_k}$. Let N be large enough that it is greater than all the e_i and that the theorem holds for N. By the theorem, a k-tuple of polynomials $(\phi_1, \phi_2, \ldots, \phi_k)$ with $\deg \phi_i \leq e_i$ gives the principal parts of an element of K with poles concentrated at $P_1 P_2 \cdots P_k$ if and only if it satisfies the g conditions $\sum_{i=1}^{k} \langle \phi_i, \xi_i^{(j)} \rangle = 0$ for $j = 1, 2, \ldots, g$. The k-tuples satisfying these g conditions are a vector space over K_0 of dimension $\sum e_i - \psi$ where ψ is the number of *independent* conditions imposed by the g conditions $\sum_{i=1}^{k} \langle \phi_i, \xi_i^{(j)} \rangle = 0$. Thus ψ is g minus the dimension of the space of holomorphic differentials for which the conditions $\sum_{i=1}^{k} \langle \phi_i, \xi_i^{(j)} \rangle = 0$ impose *no* conditions on k-tuples $(\phi_1, \phi_2, \ldots, \phi_k)$ satisfying $\deg \phi_i \leq e_i$. A holomorphic differential ω is in this space if and only if in the corresponding $(\xi_1, \xi_2, \ldots, \xi_k)$ no ξ_i contains a term of degree $\leq e_i$. Therefore, $\psi = g - \mu$, where μ is as in the statement of the corollary. Thus, the k-tuples which are principal parts of elements of K divisible by A^{-1} are a vector space of dimension $\sum e_i - g + \mu = \deg_K A - g + \mu$. Two elements of K have the same principal parts if and only if their difference is a constant, so $\dim_K(A^{-1})$ is one greater than the dimension of the space of principal parts of elements divisible by A^{-1}, that is, $\dim_K(A^{-1}) = \deg_K A - g + 1 + \mu$, which is the desired formula.

§A.13 The Riemann-Roch Theorem for Reciprocals of Integral Divisors.

PROPOSITION. *Let B be an integral divisor in a function field K with $\deg_K B > 0$. The differentials ω in K divisible by B^{-1} (that is, $\mathrm{ord}_P(\omega) \geq -e$ for any place P in any numerical extension of K which divides B exactly e times) are a vector space of dimension $\deg_K B + g - 1$ over the field of constants K_0 of K.*

PROOF: Let x be a parameter of K such that $B|\{x\}$ (see §3.10), let y_1, y_2, ..., y_n be an integral basis with respect to x, and let z_1, z_2, ..., z_n be its dual basis. (Here $n = \deg\{x\} = [K : K_0(x)]$.) It was shown in §A.10 that the poles of the differential $x^i z_j dx$ are concentrated at $\{x\}$ when $i \leq e_j - 2$ (where e_j is the order of y_j at $x = \infty$). It is holomorphic when, in addition, $i \geq 0$. It is divisible by $\{x\}^{-1}$ when $x^{i+1} z_j dx$ has no pole at $\{x\}$, which is true when $i + 1 \geq 0$. Thus, the dimension of the vector space of differentials divisible by $\{x\}^{-1}$ is equal to the number of indices i for which $-1 \leq i \leq e_j - 2$ ($j = 1$, 2, ..., n). Therefore, the dimension is $g + n - 1$, because g is the number of solutions of $0 \leq i \leq e_j - 2$ and there is one additional solution $i = -1$ for each $j = 1$, 2, ..., n except for the one j for which $e_j = 0$.

Passage to a numerical extension of K does not change the dimension of the space of differentials divisible by B^{-1} or the space of differentials divisible by $\{x\}^{-1}$. Assume without loss of generality that $\{x\}$, and therefore B, is a product of places, say $\{x\} = P_1 P_2 \cdots P_n$ and $B = P_1 P_2 \cdots P_\nu$, where $1 \leq \nu \leq n$. To say that a differential ω in K which is divisible by $\{x\}^{-1}$ is also divisible by B^{-1} imposes $n - \nu$ conditions on ω, namely, the conditions that $n - \nu$ of the coefficients of the expansions of ω in terms of local parameters at the P_i are zero. (If the P_i are distinct, the conditions are simply that the residues of ω at P_i are zero for $i > \nu$.) These conditions are *independent* because the differentials divisible by $\{x\}^{-1}$ have dimension $g + n - 1$ and the differentials divisible by P_1^{-1} have dimension g (a differential divisible by P_1^{-1} has no poles except, possibly, for a simple pole at P_1, but, because the sum of the residues is zero, it can also have no pole at P_1 and must therefore be holomorphic) so the $n-1$ conditions imposed by divisibility by P_1^{-1} must be independent. Therefore, the differentials divisible by $B^{-1} = (P_1 P_2 \cdots P_\nu)^{-1}$ have dimension $g + n - 1 - (n - \nu) = g + \nu - 1$, as was to be shown.

COROLLARY. *The Riemann-Roch formula*

$$\dim_K A^{-1} + \deg_K A^{-1} + g = 1 + \mu,$$

where μ is the dimension of the space of differentials divisible by A, holds when A is the reciprocal of an integral divisor.

DEDUCTION: If $A = [1]$, the formula was proved in §A.12. Otherwise, $A = B^{-1}$ where B is as in the Proposition. Then $\dim_K A^{-1} = \dim_K B = 0$ because a nonzero element z of K divisible by B would satisfy $\deg_K[z] \geq 1$. Thus $\dim_K A^{-1} + \deg_K A^{-1} + g = 0 + \nu + g = 1 + \mu$, where $\nu = \deg_K B$ and μ is the dimension $g + \nu - 1$ of the space of differentials divisible by $A = B^{-1}$.

§A.14 General Case of the Riemann-Roch Theorem.

Again, let P_1, P_2, ..., P_k be places in a function field K and let u_i be a local parameter at P_i for each i. Let B be an integral divisor with $\deg_K B = \nu > 0$ which is relatively prime to $P_1 P_2 \cdots P_k$, and let $\omega_1, \omega_2, \ldots, \omega_{g+\nu-1}$ be a basis of the vector space of differentials in K divisible by B^{-1}. The principal parts $(\phi_1, \phi_2, \ldots, \phi_k)$ of any element z of K which has poles concentrated at $P_1 P_2 \cdots P_k$ and which is zero at B

satisfy $\sum_{i=1}^{k} \langle \phi_i, \xi_i^{(j)} \rangle = 0$ for $j = 1, 2, \ldots, g + \nu - 1$, where the $\xi_i^{(j)}$ are defined by $\omega_j = (\xi_i^{(j)}(u_i) + o(u_i^N)) du_i$ for some N larger than the degrees of the ϕ_i, and where $\langle \phi_i, \xi_i^{(j)} \rangle$ is the coefficient of X^{-1} in $\phi_i(X^{-1}) \xi_i^{(j)}(X)$. *These necessary conditions on* $(\phi_1, \phi_2, \ldots, \phi_k)$ *are also sufficient, that is, given a k-tuple of polynomials* $(\phi_1, \phi_2, \ldots, \phi_k)$ *(coefficients in K_0, zero constant terms) satisfying the $g + \nu - 1$ conditions* $\sum \langle \phi_i, \xi_i^{(j)} \rangle = 0$*, there is an element z of K zero at B and with poles concentrated at $P_1 P_2 \cdots P_k$ whose principal parts are the* ϕ_i*.*

PROOF: Let $A = P_1 P_2 \cdots P_k$. For N sufficiently large, $\dim_K(A^{-N}B) + \deg_K(A^{-N}B) + g = 1$ (§3.26). That is, $\dim_K(A^{-N}B) = Nk - \nu - g + 1$. Since an element of K divisible by $A^{-N}B$ is determined by its principal parts (the difference of two with the same principal parts has no poles, therefore is constant, therefore is zero because it is zero at B) it follows that the set of principal parts has codimension $\nu + g - 1$ in the Nk-dimensional space of k-tuples $(\phi_1, \phi_2, \ldots, \phi_k)$ in which $\deg \phi_i \leq N$. Since, for N sufficiently large, the $\nu + g - 1$ necessary conditions $\sum \langle \phi_i, \xi_i^{(j)} \rangle = 0$ are independent, it follows as in §A.12 that they suffice to determine the k-tuples that occur as principal parts.

COROLLARY. (Riemann-Roch Theorem) *For any divisor C in a function field K, $\dim_K(C^{-1}) + \deg_K(C^{-1}) + g = 1 + \mu$, where μ is the dimension of the space of differentials in K divisible by C (as a vector space over the field of constants K_0 of K).*

DEDUCTION: One can assume without loss of generality that C is a product of powers of places (§3.23), say $C = A/B$ where $A = P_1^{e_1} P_2^{e_2} \cdots P_k^{e_k}$ and $B = P_{k+1}^{e_{k+1}} \cdots P_t^{e_t}$ for distinct places P_i and for positive integers e_i. Also, because the other cases have already been proved, one can assume $\deg_K A > 0$ and $\deg_K B > 0$. Then $\dim_K(C^{-1})$ is the dimension of the space of all k-tuples $(\phi_1, \phi_2, \ldots, \phi_k)$ of principal parts of elements z of K which are zero at B, have poles concentrated at $P_1 P_2 \cdots P_k$, and have principal parts satisfying $\deg \phi_i < e_i$. This dimension is $e_1 + e_2 + \cdots + e_k - \psi = \deg_K A - \psi$, where ψ is the number of *independent* conditions imposed by the $\deg_K B + g - 1$ conditions $\sum \langle \phi_i, \xi_i^{(j)} \rangle = 0$ ($j = 1, 2, \ldots,$ $\deg_K B + g - 1$). Since $\psi = \deg_K B + g - 1 - \mu$ where μ is the dimension of the subspace of the differentials in K divisible by B^{-1} which are zero at A—that is, where μ is as in the Corollary—it follows that $\dim_K(C^{-1}) = \deg_K A - \deg_K B - g + 1 + \mu = \deg_K C - g + 1 + \mu$, as was to be shown.

References

[A] Abel, Niels Henrik, *Mémoire sur une propriété générale d'une classe très-étendue de fonctions transcendantes*, Présenté à l'Académie des sciences à Paris le 30 Octobre 1826, Mémoires présentés par divers savants, vol. VII, Paris, 1841. (*Oeuvres*, vol. 1, 145-211.)

[D1] Dedekind, Richard, *Ueber einen arithmetischen Satz von Gauss*, Mitt. Deut. Math. Gesell. Prague, 1892 1-11. (*Werke*, vol. 2, 28-38.)

[D2] Dedekind, Richard, *Ueber die Begründung der Idealtheorie*, Nachr. Kön. Ges. Wiss. Göttingen, 1895, 106-113. (*Werke*, vol. 2, 50-58.)

[D-W] Dedekind, Richard and Weber, Heinrich, *Theorie der algebraischen Funktionen einer Veränderlichen*, Jour. für Math., 92 (1882) 181-290. (Dedekind's *Werke*, 1, 238-349.)

[E1] Edwards, Harold M., *The Genesis of Ideal Theory*, Arch. Hist. Ex. Sci., 23 (1980) 321-378.

[E2] Edwards, Harold M., *Dedekind's Invention of Ideals*, Bull. Lond. Math. Soc., 15 (1983) 8-17. (*Studies in the History of Mathematics*, Esther R. Phillips, ed., MAA, 1987, 8-20.)

[E3] Edwards, Harold M., Galois Theory. Springer-Verlag, New York Berlin Heidelberg Tokyo, 1984.

[E-N-P] Edwards, Harold; Neumann, Olaf; and Purkert, Walter, *Dedekinds "Bunte Bemerkungen" zu Kroneckers "Grundzüge"*, Arch. Hist. Ex. Sci., 27 (1982) 49-85.

[He] Hensel, Kurt, *Untersuchung der Fundamentalgleichung einer Gattung für eine reelle Primzahl als Modul und Bestimmung der Theiler ihrer Discriminante*, Jour. für Math. 113 (1894) 61-83.

[Hu1] Hurwitz, Adolph, *Ueber die Theorie der Ideale*, Nachr. kön. Ges. Wiss. Göttingen, 1894, 291-298. (*Werke*, vol. 2, 191-197.)

[Hu2] Hurwitz, Adolph, *Ueber einen Fundamentalsatz der arithmetischen Theorie der algebraischen Größen*, Nachr. kön

Ges. Wiss. Göttingen, 1895, 230-240. (*Werke*, vol. 2, 198-207.)

[Ko] König, Julius, Einleitung in die allgemeine Theorie der algebraischen Grössen. Teubner, Leipzig, 1903.

[Kr1] Kronecker, Leopold, *Grundzüge einer arithmetischen Theorie der algebraischen Grössen*, Jour. für Math. 92 (1882) 1-122. (Also published separately by Reimer, Berlin, 1882. Also *Werke*, vol. 2, 237-387.)

[Kr2] Kronecker, Leopold, *Zur Theorie der Formen höherer Stufen*, Monatsber. Akad. Wiss. Berlin, 1883, 957-960. (Also *Werke*, vol. 2, 419-424.)

[Ku] Kummer, Ernst Eduard, *Zur Theorie der Complexen Zahlen*, Monatsber. Akad Wiss. Berlin, 1846, 87-96. (Also, Jour. für Math., 35 (1847) 319-326 and Collected Papers, vol. 1, 203-210).

[Wa] Walker, Robert J., Algebraic Curves, Princeton University Press, Princeton, 1950. (Reprint, Dover, New York, 1962.)

[Wb] Weber, Heinrich, Lehrbuch der Algebra, Vieweg, Braunschweig, 1895, 1896; reprint Chelsea, New York, 1962, 1979.

[We] Weyl, Hermann, Algebraic Theory of Numbers. Princeton University Press, Princeton, 1940, 1951.

Printed in the United States
By Bookmasters